Sustainability: A Very Short Introduction

VERY SHORT INTRODUCTIONS are for anyone wanting a stimulating and accessible way into a new subject. They are written by experts, and have been translated into more than 45 different languages.

The series began in 1995, and now covers a wide variety of topics in every discipline. The VSI library currently contains over 750 volumes—a Very Short Introduction to everything from Psychology and Philosophy of Science to American History and Relativity—and continues to grow in every subject area.

Very Short Introductions available now:

ABOLITIONISM Richard S. Newman
THE ABRAHAMIC RELIGIONS Charles L. Cohen
ACCOUNTING Christopher Nobes
ADDICTION Keith Humphreys
ADOLESCENCE Peter K. Smith
THEODOR W. ADORNO Andrew Bowie
ADVERTISING Winston Fletcher
AERIAL WARFARE Frank Ledwidge
AESTHETICS Bence Nanay
AFRICAN AMERICAN HISTORY Jonathan Scott Holloway
AFRICAN AMERICAN RELIGION Eddie S. Glaude Jr.
AFRICAN HISTORY John Parker and Richard Rathbone
AFRICAN POLITICS Ian Taylor
AFRICAN RELIGIONS Jacob K. Olupona
AGEING Nancy A. Pachana
AGNOSTICISM Robin Le Poidevin
AGRICULTURE Paul Brassley and Richard Soffe
ALEXANDER THE GREAT Hugh Bowden
ALGEBRA Peter M. Higgins
AMERICAN BUSINESS HISTORY Walter A. Friedman
AMERICAN CULTURAL HISTORY Eric Avila
AMERICAN FOREIGN RELATIONS Andrew Preston
AMERICAN HISTORY Paul S. Boyer
AMERICAN IMMIGRATION David A. Gerber
AMERICAN INTELLECTUAL HISTORY Jennifer Ratner-Rosenhagen
THE AMERICAN JUDICIAL SYSTEM Charles L. Zelden
AMERICAN LEGAL HISTORY G. Edward White
AMERICAN MILITARY HISTORY Joseph T. Glatthaar
AMERICAN NAVAL HISTORY Craig L. Symonds
AMERICAN POETRY David Caplan
AMERICAN POLITICAL HISTORY Donald Critchlow
AMERICAN POLITICAL PARTIES AND ELECTIONS L. Sandy Maisel
AMERICAN POLITICS Richard M. Valelly
THE AMERICAN PRESIDENCY Charles O. Jones
THE AMERICAN REVOLUTION Robert J. Allison
AMERICAN SLAVERY Heather Andrea Williams
THE AMERICAN SOUTH Charles Reagan Wilson
THE AMERICAN WEST Stephen Aron
AMERICAN WOMEN'S HISTORY Susan Ware
AMPHIBIANS T. S. Kemp
ANAESTHESIA Aidan O'Donnell

ANALYTIC PHILOSOPHY Michael Beaney
ANARCHISM Alex Prichard
ANCIENT ASSYRIA Karen Radner
ANCIENT EGYPT Ian Shaw
ANCIENT EGYPTIAN ART AND ARCHITECTURE Christina Riggs
ANCIENT GREECE Paul Cartledge
ANCIENT GREEK AND ROMAN SCIENCE Liba Taub
THE ANCIENT NEAR EAST Amanda H. Podany
ANCIENT PHILOSOPHY Julia Annas
ANCIENT WARFARE Harry Sidebottom
ANGELS David Albert Jones
ANGLICANISM Mark Chapman
THE ANGLO-SAXON AGE John Blair
ANIMAL BEHAVIOUR Tristram D. Wyatt
THE ANIMAL KINGDOM Peter Holland
ANIMAL RIGHTS David DeGrazia
ANSELM Thomas Williams
THE ANTARCTIC Klaus Dodds
ANTHROPOCENE Erle C. Ellis
ANTISEMITISM Steven Beller
ANXIETY Daniel Freeman and Jason Freeman
THE APOCRYPHAL GOSPELS Paul Foster
APPLIED MATHEMATICS Alain Goriely
THOMAS AQUINAS Fergus Kerr
ARBITRATION Thomas Schultz and Thomas Grant
ARCHAEOLOGY Paul Bahn
ARCHITECTURE Andrew Ballantyne
THE ARCTIC Klaus Dodds and Jamie Woodward
HANNAH ARENDT Dana Villa
ARISTOCRACY William Doyle
ARISTOTLE Jonathan Barnes
ART HISTORY Dana Arnold
ART THEORY Cynthia Freeland
ARTIFICIAL INTELLIGENCE Margaret A. Boden
ASIAN AMERICAN HISTORY Madeline Y. Hsu
ASTROBIOLOGY David C. Catling
ASTROPHYSICS James Binney
ATHEISM Julian Baggini
THE ATMOSPHERE Paul I. Palmer
AUGUSTINE Henry Chadwick
JANE AUSTEN Tom Keymer
AUSTRALIA Kenneth Morgan
AUTHORITARIANISM James Loxton
AUTISM Uta Frith
AUTOBIOGRAPHY Laura Marcus
THE AVANT GARDE David Cottington
THE AZTECS Davíd Carrasco
BABYLONIA Trevor Bryce
BACTERIA Sebastian G. B. Amyes
BANKING John Goddard and John O. S. Wilson
BARTHES Jonathan Culler
THE BEATS David Sterritt
BEAUTY Roger Scruton
LUDWIG VAN BEETHOVEN Mark Evan Bonds
BEHAVIOURAL ECONOMICS Michelle Baddeley
BESTSELLERS John Sutherland
THE BIBLE John Riches
BIBLICAL ARCHAEOLOGY Eric H. Cline
BIG DATA Dawn E. Holmes
BIOCHEMISTRY Mark Lorch
BIODIVERSITY CONSERVATION David Macdonald
BIOGEOGRAPHY Mark V. Lomolino
BIOGRAPHY Hermione Lee
BIOMETRICS Michael Fairhurst
ELIZABETH BISHOP Jonathan F. S. Post
BLACK HOLES Katherine Blundell
BLASPHEMY Yvonne Sherwood
BLOOD Chris Cooper
THE BLUES Elijah Wald
THE BODY Chris Shilling
THE BOHEMIANS David Weir
NIELS BOHR J. L. Heilbron
THE BOOK OF COMMON PRAYER Brian Cummings
THE BOOK OF MORMON Terryl Givens
BORDERS Alexander C. Diener and Joshua Hagen
THE BRAIN Michael O'Shea
BRANDING Robert Jones
THE BRICS Andrew F. Cooper
BRITISH ARCHITECTURE Dana Arnold

BRITISH CINEMA Charles Barr
THE BRITISH CONSTITUTION Martin Loughlin
THE BRITISH EMPIRE Ashley Jackson
BRITISH POLITICS Tony Wright
BUDDHA Michael Carrithers
BUDDHISM Damien Keown
BUDDHIST ETHICS Damien Keown
BYZANTIUM Peter Sarris
CALVINISM Jon Balserak
ALBERT CAMUS Oliver Gloag
CANADA Donald Wright
CANCER Nicholas James
CAPITALISM James Fulcher
CATHOLICISM Gerald O'Collins
CAUSATION Stephen Mumford and Rani Lill Anjum
THE CELL Terence Allen and Graham Cowling
THE CELTS Barry Cunliffe
CHAOS Leonard Smith
GEOFFREY CHAUCER David Wallace
CHEMISTRY Peter Atkins
CHILD PSYCHOLOGY Usha Goswami
CHILDREN'S LITERATURE Kimberley Reynolds
CHINESE LITERATURE Sabina Knight
CHOICE THEORY Michael Allingham
CHRISTIAN ART Beth Williamson
CHRISTIAN ETHICS D. Stephen Long
CHRISTIANITY Linda Woodhead
CIRCADIAN RHYTHMS Russell Foster and Leon Kreitzman
CITIZENSHIP Richard Bellamy
CITY PLANNING Carl Abbott
CIVIL ENGINEERING David Muir Wood
THE CIVIL RIGHTS MOVEMENT Thomas C. Holt
CIVIL WARS Monica Duffy Toft
CLASSICAL LITERATURE William Allan
CLASSICAL MYTHOLOGY Helen Morales
CLASSICS Mary Beard and John Henderson
CLAUSEWITZ Michael Howard
CLIMATE Mark Maslin
CLIMATE CHANGE Mark Maslin
CLINICAL PSYCHOLOGY Susan Llewelyn and Katie Aafjes-van Doorn
COGNITIVE BEHAVIOURAL THERAPY Freda McManus
COGNITIVE NEUROSCIENCE Richard Passingham
THE COLD WAR Robert J. McMahon
COLONIAL AMERICA Alan Taylor
COLONIAL LATIN AMERICAN LITERATURE Rolena Adorno
COMBINATORICS Robin Wilson
COMEDY Matthew Bevis
COMMUNISM Leslie Holmes
COMPARATIVE LAW Sabrina Ragone and Guido Smorto
COMPARATIVE LITERATURE Ben Hutchinson
COMPETITION AND ANTITRUST LAW Ariel Ezrachi
COMPLEXITY John H. Holland
THE COMPUTER Darrel Ince
COMPUTER SCIENCE Subrata Dasgupta
CONCENTRATION CAMPS Dan Stone
CONDENSED MATTER PHYSICS Ross H. McKenzie
CONFUCIANISM Daniel K. Gardner
THE CONQUISTADORS Matthew Restall and Felipe Fernández-Armesto
CONSCIENCE Paul Strohm
CONSCIOUSNESS Susan Blackmore
CONTEMPORARY ART Julian Stallabrass
CONTEMPORARY FICTION Robert Eaglestone
CONTINENTAL PHILOSOPHY Simon Critchley
COPERNICUS Owen Gingerich
CORAL REEFS Charles Sheppard
CORPORATE SOCIAL RESPONSIBILITY Jeremy Moon
CORRUPTION Leslie Holmes
COSMOLOGY Peter Coles
COUNTRY MUSIC Richard Carlin
CREATIVITY Vlad Glăveanu
CRIME FICTION Richard Bradford
CRIMINAL JUSTICE Julian V. Roberts
CRIMINOLOGY Tim Newburn
CRITICAL THEORY Stephen Eric Bronner
THE CRUSADES Christopher Tyerman

CRYPTOGRAPHY Fred Piper and Sean Murphy
CRYSTALLOGRAPHY A. M. Glazer
THE CULTURAL REVOLUTION Richard Curt Kraus
DADA AND SURREALISM David Hopkins
DANTE Peter Hainsworth and David Robey
DARWIN Jonathan Howard
THE DEAD SEA SCROLLS Timothy H. Lim
DECADENCE David Weir
DECOLONIZATION Dane Kennedy
DEMENTIA Kathleen Taylor
DEMOCRACY Naomi Zack
DEMOGRAPHY Sarah Harper
DEPRESSION Jan Scott and Mary Jane Tacchi
DERRIDA Simon Glendinning
DESCARTES Tom Sorell
DESERTS Nick Middleton
DESIGN John Heskett
DEVELOPMENT Ian Goldin
DEVELOPMENTAL BIOLOGY Lewis Wolpert
THE DEVIL Darren Oldridge
DIASPORA Kevin Kenny
CHARLES DICKENS Jenny Hartley
DICTIONARIES Lynda Mugglestone
DINOSAURS David Norman
DIPLOMATIC HISTORY Joseph M. Siracusa
DOCUMENTARY FILM Patricia Aufderheide
DOSTOEVSKY Deborah Martinsen
DREAMING J. Allan Hobson
DRUGS Les Iversen
DRUIDS Barry Cunliffe
DYNASTY Jeroen Duindam
DYSLEXIA Margaret J. Snowling
EARLY MUSIC Thomas Forrest Kelly
THE EARTH Martin Redfern
EARTH SYSTEM SCIENCE Tim Lenton
ECOLOGY Jaboury Ghazoul
ECONOMICS Partha Dasgupta
EDUCATION Gary Thomas
EGYPTIAN MYTH Geraldine Pinch
EIGHTEENTH-CENTURY BRITAIN Paul Langford
ELECTIONS L. Sandy Maisel and Jennifer A. Yoder
THE ELEMENTS Philip Ball
EMOTION Dylan Evans
EMPIRE Stephen Howe
EMPLOYMENT LAW David Cabrelli
ENERGY SYSTEMS Nick Jenkins
ENGELS Terrell Carver
ENGINEERING David Blockley
THE ENGLISH LANGUAGE Simon Horobin
ENGLISH LITERATURE Jonathan Bate
THE ENLIGHTENMENT John Robertson
ENTREPRENEURSHIP Paul Westhead and Mike Wright
ENVIRONMENTAL ECONOMICS Stephen Smith
ENVIRONMENTAL ETHICS Robin Attfield
ENVIRONMENTAL LAW Elizabeth Fisher
ENVIRONMENTAL POLITICS Andrew Dobson
ENZYMES Paul Engel
THE EPIC Anthony Welch
EPICUREANISM Catherine Wilson
EPIDEMIOLOGY Rodolfo Saracci
ETHICS Simon Blackburn
ETHNOMUSICOLOGY Timothy Rice
THE ETRUSCANS Christopher Smith
EUGENICS Philippa Levine
THE EUROPEAN UNION Simon Usherwood and John Pinder
EUROPEAN UNION LAW Anthony Arnull
EVANGELICALISM John G. Stackhouse Jr.
EVIL Luke Russell
EVOLUTION Brian and Deborah Charlesworth
EXISTENTIALISM Thomas Flynn
EXPLORATION Stewart A. Weaver
EXTINCTION Paul B. Wignall
THE EYE Michael Land
FAIRY TALE Marina Warner
FAITH Roger Trigg
FAMILY LAW Jonathan Herring
MICHAEL FARADAY Frank A. J. L. James

FASCISM Kevin Passmore
FASHION Rebecca Arnold
FEDERALISM Mark J. Rozell and Clyde Wilcox
FEMINISM Margaret Walters
FEMINIST PHILOSOPHY Katharine Jenkins
FILM Michael Wood
FILM MUSIC Kathryn Kalinak
FILM NOIR James Naremore
FIRE Andrew C. Scott
THE FIRST WORLD WAR Michael Howard
FLUID MECHANICS Eric Lauga
FOLK MUSIC Mark Slobin
FOOD John Krebs
FORENSIC PSYCHOLOGY David Canter
FORENSIC SCIENCE Jim Fraser
FORESTS Jaboury Ghazoul
FOSSILS Keith Thomson
FOUCAULT Gary Gutting
THE FOUNDING FATHERS R. B. Bernstein
FRACTALS Kenneth Falconer
FREE SPEECH Nigel Warburton
FREE WILL Thomas Pink
FREEMASONRY Andreas Önnerfors
FRENCH CINEMA Dudley Andrew
FRENCH LITERATURE John D. Lyons
FRENCH PHILOSOPHY Stephen Gaukroger and Knox Peden
THE FRENCH REVOLUTION William Doyle
FREUD Anthony Storr
FUNDAMENTALISM Malise Ruthven
FUNGI Nicholas P. Money
THE FUTURE Jennifer M. Gidley
GALAXIES John Gribbin
GALILEO Stillman Drake
GAME THEORY Ken Binmore
GANDHI Bhikhu Parekh
GARDEN HISTORY Gordon Campbell
GENDER HISTORY Antoinette Burton
GENES Jonathan Slack
GENIUS Andrew Robinson
GENOMICS John Archibald
GEOGRAPHY John Matthews and David Herbert
GEOLOGY Jan Zalasiewicz
GEOMETRY Maciej Dunajski
GEOPHYSICAL AND CLIMATE HAZARDS Bill McGuire
GEOPHYSICS William Lowrie
GEOPOLITICS Klaus Dodds
GERMAN LITERATURE Nicholas Boyle
GERMAN PHILOSOPHY Andrew Bowie
THE GHETTO Bryan Cheyette
GLACIATION David J. A. Evans
GLOBAL ECONOMIC HISTORY Robert C. Allen
GLOBAL ISLAM Nile Green
GLOBALIZATION Manfred B. Steger
GOD John Bowker
GÖDEL'S THEOREM A. W. Moore
GOETHE Ritchie Robertson
THE GOTHIC Nick Groom
GOVERNANCE Mark Bevir
GRAVITY Timothy Clifton
THE GREAT DEPRESSION AND THE NEW DEAL Eric Rauchway
THE GULAG Alan Barenberg
HABEAS CORPUS Amanda L. Tyler
HABERMAS James Gordon Finlayson
THE HABSBURG EMPIRE Martyn Rady
HAPPINESS Daniel M. Haybron
THE HARLEM RENAISSANCE Cheryl A. Wall
THE HEBREW BIBLE AS LITERATURE Tod Linafelt
HEGEL Peter Singer
HEIDEGGER Michael Inwood
THE HELLENISTIC AGE Peter Thonemann
HEREDITY John Waller
HERMENEUTICS Jens Zimmermann
HERODOTUS Jennifer T. Roberts
HIEROGLYPHS Penelope Wilson
HINDUISM Kim Knott
HISTORY John H. Arnold
THE HISTORY OF ASTRONOMY Michael Hoskin
THE HISTORY OF CHEMISTRY William H. Brock
THE HISTORY OF CHILDHOOD James Marten
THE HISTORY OF CINEMA Geoffrey Nowell-Smith

THE HISTORY OF COMPUTING
Doron Swade
THE HISTORY OF EMOTIONS
Thomas Dixon
THE HISTORY OF LIFE Michael Benton
THE HISTORY OF MATHEMATICS
Jacqueline Stedall
THE HISTORY OF MEDICINE
William Bynum
THE HISTORY OF PHYSICS
J. L. Heilbron
THE HISTORY OF POLITICAL THOUGHT Richard Whatmore
THE HISTORY OF TIME
Leofranc Holford-Strevens
HIV AND AIDS Alan Whiteside
HOBBES Richard Tuck
HOLLYWOOD Peter Decherney
THE HOLY ROMAN EMPIRE
Joachim Whaley
HOME Michael Allen Fox
HOMER Barbara Graziosi
HORACE Llewelyn Morgan
HORMONES Martin Luck
HORROR Darryl Jones
HUMAN ANATOMY Leslie Klenerman
HUMAN EVOLUTION Bernard Wood
HUMAN PHYSIOLOGY
Jamie A. Davies
HUMAN RESOURCE MANAGEMENT Adrian Wilkinson
HUMAN RIGHTS Andrew Clapham
HUMANISM Stephen Law
HUME James A. Harris
HUMOUR Noël Carroll
IBN SĪNĀ (AVICENNA) Peter Adamson
THE ICE AGE Jamie Woodward
IDENTITY Florian Coulmas
IDEOLOGY Michael Freeden
IMAGINATION
Jennifer Gosetti-Ferencei
THE IMMUNE SYSTEM
Paul Klenerman
INDIAN CINEMA Ashish Rajadhyaksha
INDIAN PHILOSOPHY Sue Hamilton
THE INDUSTRIAL REVOLUTION
Robert C. Allen
INFECTIOUS DISEASE
Marta L. Wayne and Benjamin M. Bolker
INFINITY Ian Stewart
INFORMATION Luciano Floridi
INNOVATION Mark Dodgson and David Gann
INTELLECTUAL PROPERTY
Siva Vaidhyanathan
INTELLIGENCE Ian J. Deary
INTERNATIONAL LAW
Vaughan Lowe
INTERNATIONAL MIGRATION
Khalid Koser
INTERNATIONAL RELATIONS
Christian Reus-Smit
INTERNATIONAL SECURITY
Christopher S. Browning
INSECTS Simon Leather
INVASIVE SPECIES Julie Lockwood and Dustin Welbourne
IRAN Ali M. Ansari
ISLAM Malise Ruthven
ISLAMIC HISTORY Adam Silverstein
ISLAMIC LAW Mashood A. Baderin
ISOTOPES Rob Ellam
ITALIAN LITERATURE
Peter Hainsworth and David Robey
HENRY JAMES Susan L. Mizruchi
JAPANESE LITERATURE Alan Tansman
JESUS Richard Bauckham
JEWISH HISTORY David N. Myers
JEWISH LITERATURE Ilan Stavans
JOURNALISM Ian Hargreaves
JAMES JOYCE Colin MacCabe
JUDAISM Norman Solomon
JUNG Anthony Stevens
THE JURY Renée Lettow Lerner
KABBALAH Joseph Dan
KAFKA Ritchie Robertson
KANT Roger Scruton
KEYNES Robert Skidelsky
KIERKEGAARD Patrick Gardiner
KNOWLEDGE Jennifer Nagel
THE KORAN Michael Cook
KOREA Michael J. Seth
LAKES Warwick F. Vincent
LANDSCAPE ARCHITECTURE
Ian H. Thompson
LANDSCAPES AND GEOMORPHOLOGY
Andrew Goudie and Heather Viles
LANGUAGES Stephen R. Anderson

LATE ANTIQUITY Gillian Clark
LAW Raymond Wacks
THE LAWS OF THERMODYNAMICS Peter Atkins
LEADERSHIP Keith Grint
LEARNING Mark Haselgrove
LEIBNIZ Maria Rosa Antognazza
C. S. LEWIS James Como
LIBERALISM Michael Freeden
LIGHT Ian Walmsley
LINCOLN Allen C. Guelzo
LINGUISTICS Peter Matthews
LITERARY THEORY Jonathan Culler
LOCKE John Dunn
LOGIC Graham Priest
LOVE Ronald de Sousa
MARTIN LUTHER Scott H. Hendrix
MACHIAVELLI Quentin Skinner
MADNESS Andrew Scull
MAGIC Owen Davies
MAGNA CARTA Nicholas Vincent
MAGNETISM Stephen Blundell
MALTHUS Donald Winch
MAMMALS T. S. Kemp
MANAGEMENT John Hendry
NELSON MANDELA Elleke Boehmer
MAO Delia Davin
MARINE BIOLOGY Philip V. Mladenov
MARKETING Kenneth Le Meunier-FitzHugh
THE MARQUIS DE SADE John Phillips
MARTYRDOM Jolyon Mitchell
MARX Peter Singer
MATERIALS Christopher Hall
MATHEMATICAL ANALYSIS Richard Earl
MATHEMATICAL FINANCE Mark H. A. Davis
MATHEMATICS Timothy Gowers
MATTER Geoff Cottrell
THE MAYA Matthew Restall and Amara Solari
THE MEANING OF LIFE Terry Eagleton
MEASUREMENT David Hand
MEDICAL ETHICS Michael Dunn andTony Hope
MEDICAL LAW Charles Foster
MEDIEVAL BRITAIN John Gillingham and Ralph A. Griffiths
MEDIEVAL LITERATURE Elaine Treharne
MEDIEVAL PHILOSOPHY John Marenbon
MEMORY Jonathan K. Foster
METAPHYSICS Stephen Mumford
METHODISM William J. Abraham
THE MEXICAN REVOLUTION Alan Knight
MICROBIOLOGY Nicholas P. Money
MICROBIOMES Angela E. Douglas
MICROECONOMICS Avinash Dixit
MICROSCOPY Terence Allen
THE MIDDLE AGES Miri Rubin
MILITARY JUSTICE Eugene R. Fidell
MILITARY STRATEGY Antulio J. Echevarria II
JOHN STUART MILL Gregory Claeys
MINERALS David Vaughan
MIRACLES Yujin Nagasawa
MODERN ARCHITECTURE Adam Sharr
MODERN ART David Cottington
MODERN BRAZIL Anthony W. Pereira
MODERN CHINA Rana Mitter
MODERN DRAMA Kirsten E. Shepherd-Barr
MODERN FRANCE Vanessa R. Schwartz
MODERN INDIA Craig Jeffrey
MODERN IRELAND Senia Pašeta
MODERN ITALY Anna Cento Bull
MODERN JAPAN Christopher Goto-Jones
MODERN LATIN AMERICAN LITERATURE Roberto González Echevarría
MODERN WAR Richard English
MODERNISM Christopher Butler
MOLECULAR BIOLOGY Aysha Divan and Janice A. Royds
MOLECULES Philip Ball
MONASTICISM Stephen J. Davis
THE MONGOLS Morris Rossabi
MONTAIGNE William M. Hamlin
MOONS David A. Rothery
MORMONISM Richard Lyman Bushman
MOUNTAINS Martin F. Price
MUHAMMAD Jonathan A. C. Brown

MULTICULTURALISM Ali Rattansi
MULTILINGUALISM John C. Maher
MUSIC Nicholas Cook
MUSIC AND TECHNOLOGY Mark Katz
MYTH Robert A. Segal
NANOTECHNOLOGY Philip Moriarty
NAPOLEON David A. Bell
THE NAPOLEONIC WARS Mike Rapport
NATIONALISM Steven Grosby
NATIVE AMERICAN LITERATURE Sean Teuton
NAVIGATION Jim Bennett
NAZI GERMANY Jane Caplan
NEGOTIATION Carrie Menkel-Meadow
NEOLIBERALISM Manfred B. Steger and Ravi K. Roy
NETWORKS Guido Caldarelli and Michele Catanzaro
THE NEW TESTAMENT Luke Timothy Johnson
THE NEW TESTAMENT AS LITERATURE Kyle Keefer
NEWTON Robert Iliffe
NIETZSCHE Michael Tanner
NINETEENTH-CENTURY BRITAIN Christopher Harvie and H. C. G. Matthew
THE NORMAN CONQUEST George Garnett
NORTH AMERICAN INDIANS Theda Perdue and Michael D. Green
NORTHERN IRELAND Marc Mulholland
NOTHING Frank Close
NUCLEAR PHYSICS Frank Close
NUCLEAR POWER Maxwell Irvine
NUCLEAR WEAPONS Joseph M. Siracusa
NUMBER THEORY Robin Wilson
NUMBERS Peter M. Higgins
NUTRITION David A. Bender
OBJECTIVITY Stephen Gaukroger
OBSERVATIONAL ASTRONOMY Geoff Cottrell
OCEANS Dorrik Stow
THE OLD TESTAMENT Michael D. Coogan
THE ORCHESTRA D. Kern Holoman
ORGANIC CHEMISTRY Graham Patrick
ORGANIZATIONS Mary Jo Hatch
ORGANIZED CRIME Georgios A. Antonopoulos and Georgios Papanicolaou
ORTHODOX CHRISTIANITY A. Edward Siecienski
OVID Llewelyn Morgan
PAGANISM Owen Davies
PAKISTAN Pippa Virdee
THE PALESTINIAN-ISRAELI CONFLICT Martin Bunton
PANDEMICS Christian W. McMillen
PARTICLE PHYSICS Frank Close
PAUL E. P. Sanders
IVAN PAVLOV Daniel P. Todes
PEACE Oliver P. Richmond
PENTECOSTALISM William K. Kay
PERCEPTION Brian Rogers
THE PERIODIC TABLE Eric R. Scerri
PHILOSOPHICAL METHOD Timothy Williamson
PHILOSOPHY Edward Craig
PHILOSOPHY IN THE ISLAMIC WORLD Peter Adamson
PHILOSOPHY OF BIOLOGY Samir Okasha
PHILOSOPHY OF LAW Raymond Wacks
PHILOSOPHY OF MIND Barbara Gail Montero
PHILOSOPHY OF PHYSICS David Wallace
PHILOSOPHY OF SCIENCE Samir Okasha
PHILOSOPHY OF RELIGION Tim Bayne
PHOTOGRAPHY Steve Edwards
PHYSICAL CHEMISTRY Peter Atkins
PHYSICS Sidney Perkowitz
PILGRIMAGE Ian Reader
PLAGUE Paul Slack
PLANETARY SYSTEMS Raymond T. Pierrehumbert
PLANETS David A. Rothery
PLANTS Timothy Walker
PLATE TECTONICS Peter Molnar
SYLVIA PLATH Heather Clark
PLATO Julia Annas

POETRY Bernard O'Donoghue
POLITICAL PHILOSOPHY David Miller
POLITICS Kenneth Minogue
POLYGAMY Sarah M. S. Pearsall
POPULISM Cas Mudde and
Cristóbal Rovira Kaltwasser
POSTCOLONIALISM Robert J. C. Young
POSTMODERNISM Christopher Butler
POSTSTRUCTURALISM
Catherine Belsey
POVERTY Philip N. Jefferson
PREHISTORY Chris Gosden
PRESOCRATIC PHILOSOPHY
Catherine Osborne
PRIVACY Raymond Wacks
PROBABILITY John Haigh
PROGRESSIVISM Walter Nugent
PROHIBITION W. J. Rorabaugh
PROJECTS Andrew Davies
PROTESTANTISM Mark A. Noll
MARCEL PROUST Joshua Landy
PSEUDOSCIENCE Michael D. Gordin
PSYCHIATRY Tom Burns
PSYCHOANALYSIS Daniel Pick
PSYCHOLOGY Gillian Butler and
Freda McManus
PSYCHOLOGY OF MUSIC
Elizabeth Hellmuth Margulis
PSYCHOPATHY Essi Viding
PSYCHOTHERAPY Tom Burns and
Eva Burns-Lundgren
PUBLIC ADMINISTRATION
Stella Z. Theodoulou and Ravi K. Roy
PUBLIC HEALTH Virginia Berridge
PURITANISM Francis J. Bremer
THE QUAKERS Pink Dandelion
QUANTUM THEORY
John Polkinghorne
RACISM Ali Rattansi
RADIOACTIVITY Claudio Tuniz
RASTAFARI Ennis B. Edmonds
READING Belinda Jack
THE REAGAN REVOLUTION Gil Troy
REALITY Jan Westerhoff
RECONSTRUCTION Allen C. Guelzo
THE REFORMATION Peter Marshall
REFUGEES Gil Loescher
RELATIVITY Russell Stannard
RELIGION Thomas A. Tweed
RELIGION IN AMERICA Timothy Beal
THE RENAISSANCE Jerry Brotton
RENAISSANCE ART
Geraldine A. Johnson
RENEWABLE ENERGY Nick Jelley
REPTILES T. S. Kemp
REVOLUTIONS Jack A. Goldstone
RHETORIC Richard Toye
RISK Baruch Fischhoff and John Kadvany
RITUAL Barry Stephenson
RIVERS Nick Middleton
ROBOTICS Alan Winfield
ROCKS Jan Zalasiewicz
ROMAN BRITAIN Peter Salway
THE ROMAN EMPIRE
Christopher Kelly
THE ROMAN REPUBLIC
David M. Gwynn
ROMANTICISM Michael Ferber
ROUSSEAU Robert Wokler
THE RULE OF LAW Aziz Z. Huq
RUSSELL A. C. Grayling
THE RUSSIAN ECONOMY
Richard Connolly
RUSSIAN HISTORY Geoffrey Hosking
RUSSIAN LITERATURE Catriona Kelly
RUSSIAN POLITICS Brian D. Taylor
THE RUSSIAN REVOLUTION
S. A. Smith
SAINTS Simon Yarrow
SAMURAI Michael Wert
SAVANNAS Peter A. Furley
SCEPTICISM Duncan Pritchard
SCHIZOPHRENIA Chris Frith and
Eve Johnstone
SCHOPENHAUER Christopher Janaway
SCIENCE AND RELIGION
Thomas Dixon and Adam R. Shapiro
SCIENCE FICTION David Seed
THE SCIENTIFIC REVOLUTION
Lawrence M. Principe
SCOTLAND Rab Houston
SECULARISM Andrew Copson
THE SELF Marya Schechtman
SEXUAL SELECTION Marlene Zuk and
Leigh W. Simmons
SEXUALITY Véronique Mottier
WILLIAM SHAKESPEARE
Stanley Wells
SHAKESPEARE'S COMEDIES
Bart van Es

SHAKESPEARE'S SONNETS AND POEMS Jonathan F. S. Post
SHAKESPEARE'S TRAGEDIES Stanley Wells
GEORGE BERNARD SHAW Christopher Wixson
MARY SHELLEY Charlotte Gordon
THE SHORT STORY Andrew Kahn
SIKHISM Eleanor Nesbitt
SILENT FILM Donna Kornhaber
THE SILK ROAD James A. Millward
SLANG Jonathon Green
SLEEP Steven W. Lockley and Russell G. Foster
SMELL Matthew Cobb
ADAM SMITH Christopher J. Berry
SOCIAL AND CULTURAL ANTHROPOLOGY John Monaghan and Peter Just
SOCIALISM Michael Newman
SOCIAL PSYCHOLOGY Richard J. Crisp
SOCIAL SCIENCE Alexander Betts
SOCIAL WORK Sally Holland and Jonathan Scourfield
SOCIOLINGUISTICS John Edwards
SOCIOLOGY Steve Bruce
SOCRATES C. C. W. Taylor
SOFT MATTER Tom McLeish
SOUND Mike Goldsmith
SOUTHEAST ASIA James R. Rush
THE SOVIET UNION Stephen Lovell
THE SPANISH CIVIL WAR Helen Graham
SPANISH LITERATURE Jo Labanyi
THE SPARTANS Andrew J. Bayliss
SPINOZA Roger Scruton
SPIRITUALITY Philip Sheldrake
SPORT Mike Cronin
STARS Andrew King
STATISTICS David J. Hand
STEM CELLS Jonathan Slack
STOICISM Brad Inwood
STRUCTURAL ENGINEERING David Blockley
STUART BRITAIN John Morrill
SUBURBS Carl Abbott
THE SUN Philip Judge
SUPERCONDUCTIVITY Stephen Blundell
SUPERSTITION Stuart Vyse
SURVEILLANCE David Lyon
SUSTAINABILITY Saleem H. Ali
SYMMETRY Ian Stewart
SYNAESTHESIA Julia Simner
SYNTHETIC BIOLOGY Jamie A. Davies
SYSTEMS BIOLOGY Eberhard O. Voit
TAXATION Stephen Smith
TEETH Peter S. Ungar
TERRORISM Charles Townshend
THEATRE Marvin Carlson
THEOLOGY David F. Ford
THINKING AND REASONING Jonathan St B. T. Evans
THOUGHT Tim Bayne
THUCYDIDES Jennifer T. Roberts
TIBETAN BUDDHISM Matthew T. Kapstein
TIDES David George Bowers and Emyr Martyn Roberts
TIME Jenann Ismael
TOCQUEVILLE Harvey C. Mansfield
J. R. R. TOLKIEN Matthew Townend
LEO TOLSTOY Liza Knapp
TOPOLOGY Richard Earl
TRAGEDY Adrian Poole
TRANSLATION Matthew Reynolds
THE TREATY OF VERSAILLES Michael S. Neiberg
TRIGONOMETRY Glen Van Brummelen
THE TROJAN WAR Eric H. Cline
ANTHONY TROLLOPE Dinah Birch
TRUST Katherine Hawley
THE TUDORS John Guy
TWENTIETH-CENTURY BRITAIN Kenneth O. Morgan
TYPOGRAPHY Paul Luna
THE UNITED NATIONS Jussi M. Hanhimäki
UNIVERSITIES AND COLLEGES David Palfreyman and Paul Temple
THE U.S. CIVIL WAR Louis P. Masur
THE U.S. CONGRESS Donald A. Ritchie
THE U.S. CONSTITUTION David J. Bodenhamer
THE U.S. SUPREME COURT Linda Greenhouse
UTILITARIANISM Katarzyna de Lazari-Radek and Peter Singer

UTOPIANISM Lyman Tower Sargent
VATICAN II Shaun Blanchard and Stephen Bullivant
VETERINARY SCIENCE James Yeates
THE VICTORIANS Martin Hewitt
THE VIKINGS Julian D. Richards
VIOLENCE Philip Dwyer
THE VIRGIN MARY Mary Joan Winn Leith
THE VIRTUES Craig A. Boyd and Kevin Timpe
VIRUSES Dorothy H. Crawford
VOLCANOES Michael J. Branney and Jan Zalasiewicz
VOLTAIRE Nicholas Cronk
WAR AND RELIGION Jolyon Mitchell and Joshua Rey
WAR AND TECHNOLOGY Alex Roland
WATER John Finney
WAVES Mike Goldsmith
WEATHER Storm Dunlop
SIMONE WEIL A. Rebecca Rozelle-Stone
THE WELFARE STATE David Garland
WITCHCRAFT Malcolm Gaskill
WITTGENSTEIN A. C. Grayling
WORK Stephen Fineman
WORLD MUSIC Philip Bohlman
WORLD MYTHOLOGY David Leeming
THE WORLD TRADE ORGANIZATION Amrita Narlikar
WORLD WAR II Gerhard L. Weinberg
WRITING AND SCRIPT Andrew Robinson
ZIONISM Michael Stanislawski
ÉMILE ZOLA Brian Nelson

Available soon:

MEANING Emma Borg and Sarah A. Fisher
PSYCHOLINGUISTICS Ferenda Ferreria
TOLERATION Andrew Murphy
ANTHONY TROLLOPE Dinah Birch
SYMBIOSIS Nancy A. Moran

For more information visit our website

www.oup.com/vsi/

Saleem H. Ali

SUSTAINABILITY

A Very Short Introduction

Great Clarendon Street, Oxford, OX2 6DP,
United Kingdom

Oxford University Press is a department of the University of Oxford. It furthers the University's objective of excellence in research, scholarship, and education by publishing worldwide. Oxford is a registered trade mark of Oxford University Press in the UK and in certain other countries

Published in the United States of America by Oxford University Press
198 Madison Avenue, New York, NY 10016, United States of America

British Library Cataloguing in Publication Data
Data available

Library of Congress Control Number: 2024938845

ISBN 978–0–19–286962–3

Printed and bound by
CPI Group (UK) Ltd, Croydon, CR0 4YY

The manufacturer's authorised representative in the EU for product safety is Oxford University Press España S.A. of el Parque Empresarial San Fernando de Henares, Avenida de Castilla, 2 – 28830 Madrid (www.oup.es/en).

Contents

List of figures and table

Figures

Table

Modified from original table by Davies, G. R. (2013). Appraising weak and strong sustainability: Searching for a middle ground. *Consilience, 10*, Article 10. https://doi.org/10.7916/consilience.v0i10.4635

Chapter 1
Seeking sustainability

The word 'sustainability' has its origins in 'sustenance', which in its etymology connects with the Latin word *sustinere*, which in turn means to furnish with support for survival. In its contemporary usage, the term sustainability has its origins in renewable resource development, particularly forestry management. The earliest usage of the concept of 'sustainability' in the Western tradition can be traced to the word *Nachhaltigkeit* in German (which etymologically means having a long-lasting effect in both time and depth of impact), used by Hans Carl von Carlowitz, who managed mining enterprises on behalf of the Saxon court in Freiberg in the late 17th century. At the time there was considerable debate about the role of forestry versus mining in the court as well as the linkages between the two sectors. Timber from the forestry was used in constructing mining shafts and related infrastructure. Von Carlowitz wrote a book in which he formulated ideas for the 'sustainable use' of the forest. His view that only so much wood should be cut as could be regrown through planned reforestation projects laid the foundations for the concept of 'sustainability' entering conversations in Europe.

For humans, just as much as for forests, the biological aspects of replenishment of our population require key resource needs to be met. Clearly nourishment of biological needs for food and shelter allows for physical survival and has held a certain primacy in our approach to human 'needs'. We have to first be alive before realizing any other forms of 'well-being'. However, increasingly there is recognition that the traditional conception of a hierarchy of human needs that was popularized by psychologist Abraham Maslow in the 1940s is misleading. This hierarchy of needs now graces most introductory psychology courses and is often depicted as a pyramid, with physiological needs at the base, followed by safety, love, esteem, and self-actualization at the apex.

Maslow suggested that the higher needs become focused only when the lower needs have been met and the individual has overcome the baser necessities. In this view, a certain quantum of material well-being is essential before other aspects can be realized. While the delineation provided by Maslow is useful, there is now general consensus among psychologists that material well-being is inextricably linked to other factors in the hierarchy and that in comparison across populations, self-actualization may be higher among individuals whose physiological needs are comparably lower. In the long run societies with less material wealth but greater social resilience may well be more viable, as the endurance of many indigenous societies has shown. Seeking a sustainable society thus becomes more challenging than what a simple view of natural resource determinism might imply. Nevertheless, we are ultimately dependent on food, water, and the capacity of ecosystems to produce such sustenance, so considering some metrics for such processes is important.

The term 'sustainability' has a calming or consoling cadence, but its meaning is often not clear because the term is used in multiple ways, without adequate explanation. Frustrated with this confusion, Dutch researcher Julian Kirscherr published a notable peer-reviewed article in May 2022 titled *Bullshit in the Sustainability and Transitions Literature!* Kirscherr's provocative title stemmed from academic work on how the obfuscation of concepts is now culturally termed 'bullshit' as a technical term across a wide swathe of global communication, and hence is more than just an abusive epithet. He systematically developed a typology of how the term 'sustainability' is used to obfuscate rather than elucidate learning on planetary processes. Unpacking sustainability from its roots in science is thus essential to avoid any such specious outcomes.

Sustainability in common parlance connotes a sense of existential continuity that suggests a 'win-win' outlook for natural resource usage in meeting human aspirations. There is also an 'ecological' tone to the word—meaning it relates to how living beings like *Homo sapiens* interact with 'ecos' (Greek for our home—planet Earth). Hunter-gatherer societies which depended on a day-to-day basis on direct interaction with natural resources provide us with our earliest conceptualization of sustainability.

Our ancient ancestors developed an often unwritten or unspoken appreciation for the ebbs and flows of our Earth's elemental rhythms. Many indigenous traditions have notions of balance for maintaining their tribal resource base through some norms or traditions. Despite this closer connection to the pulse of the planet, traditional societies also faltered in their quest for maintaining this balance. They struggled, as do we, to maintain their own population's well-being while being stewards of the environment in which they lived.

Indigenous notions of sustainability were often connected to tribal identities and were also later misunderstood by Western scholars. For example, the Apache word for 'self' and 'earth' is the same—this may mislead some environmentalists to conclude that they have an irrevocable altruism for the Earth. However, the notion of 'earth' in Apache tradition is linked to particular sites rather than the planet more generally. While the overall impact of hunter-gatherer societies on the environment was much less than industrialized societies, they too made mistakes. Species such as the giant Moa bird in New Zealand as well as various mammal species in North America met their extinction due to what may be deemed 'unsustainable' hunting practices.

Extinction is perhaps the ultimate manifestation of unsustainable usage of a resource—although if we play the long game with this phenomenon as well there are some sustainable surprises along the way. Extinctions can create opportunities for other species to flourish, which in turn can create sustainable ecosystems. For example, if the dinosaurs had not been rendered extinct by the asteroid impact around 66 million years ago, it is quite likely that humans might not have evolved, as mammals themselves were confined to shrew-sized organisms during times of reptilian dominance. The unsustainable outcome of a catastrophic event led to the rise of humans, who evolved to redefine sustainability on their own terms. Initially, as with all animals, humans were focused on their own insular sense of survival. Yet, complex social systems developed to the point where in the 20th century futurist Buckminster Fuller suggested that we consider ourselves stewards of 'Spaceship Earth'.

The 'carrying capacity' for human activity

Shipping has been an apt metaphor for sustainability at many levels. While at sea, humans have contended with resource scarcity and security. Fresh water and food are often rationed and modern ships have means of recycling water. Sailors also need to keep track of weight of cargo and how many people they can allow on a vessel to maintain buoyancy. So it is not surprising that one of the metrics that began to be used for measuring sustainability had its origins in maritime usage. The vast oceans galvanized some of the earliest concerns about extinction of a species and global resolve to address this potential challenge. The use of whale blubber for 19th-century lighting oil led to such a massive crash of whale populations that the first international environmental agreement involved these marvellous animals.

The first use of the term 'carrying capacity' dates to a report from the US State Department to Congress in 1845 around shipping tonnage and duties based on the 'carrying capacity' of ships. Geographer Nathan Sayre has done a meticulous historical analysis of the use of the term and found that it was eventually appropriated by rangeland managers concerned about over-grazing. From there it entered the lexicon of ecologists and then demographers during the heyday of population growth anxieties in the 1950s and 1960s. Eugene Odum, who wrote one of the most influential textbooks on ecology in 1951, used the concept, and from there it gained mainstream traction in environmental discourse.

Ecologist Garrett Hardin famously wrote his essay 'The Tragedy of the Commons' with such a view of carrying capacity in mind. The essay argued that with limited resources in an open access 'commons', there is a scramble for exploitation which leads to a 'tragic' depletion of the resource before it can be replenished. While being the parent of four children himself, Hardin became a champion of draconian population control, even proposing what he called 'lifeboat ethics'. His maritime metaphor implied a limited weight vessel on water which could only float if some people were thrown overboard or refused onboard. Hardin's last book was his coda on population concerns and titled *The Ostrich Factor*—referring to humanity's inability to perceive the dangers of dense cities. In it, he called for coercive constraints on 'unqualified reproductive rights'. With such austere statements from Hardin and the concept's linkage to eugenicists such as David Starr Johnson (an ichthyologist and the first president of Stanford), carrying capacity became stigmatized. Global income inequality led to calls for social justice and consumption curtailment rather than using the population panic with reference to poorer countries.

A defining personality who broadened the debate and urged ecologists to consider human material consumption in concerns about carrying capacity was the ecologist Barry Commoner. In February 1970 *Time* magazine featured him on its cover as one of the most influential voices on environmental issues, following the publication of his book *The Closing Circle*. He engaged soon thereafter in a series of debates with population biologist Paul Ehrlich and physicist John Holdren (who was to later become Science Advisor to President Obama). These debates resulted in a heuristic equation which is now used in most fundamental conversations on planetary sustainability:

$$\text{Ecological Impact} = \text{Population} \times \text{Affluence} \times \text{Technology}$$

The equation is now simply called the IPAT equation and has been the subject of further debate and defined in various forms over the years. Population and affluence (which in general correlates with material and energy consumption) are fairly straightforward as contributing factors but technology has further dimensions. Some technologies can increase impact on the environment through resource consumption, while others can mitigate impacts through efficiency or pollution control. A revised form of the equation has also been developed using more mathematical rigour and termed 'Stochastic Impacts by Regression on Population, Affluence, and Technology' (STIRPAT). This modified version introduces the measure of 'ecological elasticity', which has more policy relevance as well.

A defining challenge of considering measurements of carrying capacity and hence any ecosystems level indicator of sustainability is the unpredictability of technology. In 2014 the US National Research Council held a series of meetings which resulted in a report titled *Can Earth's and Society's Systems Meet the Needs of 10 Billion People?* This report specifically noted the role of technology as a reason for diminishing the role of 'carrying capacity' as a metric. The report notes that 'carrying capacity has been "largely abandoned" in the social and policy sciences'. And further, that 'because the human species manipulates and converts its habitat and can counter the natural limits on its population, the conceptual basis of carrying capacity breaks down when considering people'. As with the IPAT equation, the suggestion made by the council is to only use it for heuristic purposes and reject attempts to calculate the capacity per se. However, with the development of a range of geospatial data sets and improved computational abilities of system dynamics modelling, there have still been ongoing efforts to try to consider ways of engaging with the concept of carrying capacity.

Around the same time as the publication of the NRC report, a team of researchers at the University of Maryland and the University of Minnesota in the United States put forward a quantitative model to consider key factors which could lead to the rise and fall of societies over time. The study gained currency as it was funded under the auspices of the National Socioecological Synthesis Center of the US National Science Foundation and some of the tools used were also under a grant from NASA. The 'Human and Nature Dynamics' (HANDY) model was quickly picked up by the media as validation of the 'perfect storm' of social and ecological factors which could lead to civilizational collapse. Dystopian Hollywood films such as *Blade Runner* or *Water World* came to be seen as realistic scenarios under specific conditions of this model. Yet the controversy surrounding its predictions was so intense that NASA had to issue a disclaimer on their website and insist that the journal which published the study also add a statement of non-endorsement at the end of the manuscript.

A major feature of the HANDY model was its linkage of environmental variables to a key social variable—income inequality. Figure 1 shows some of the broad output features of the HANDY model as it relates to the profile of societal norms.

The HANDY study had two fundamental conclusions which should not seem so controversial but were perhaps unsettling to either end of the political spectrum:

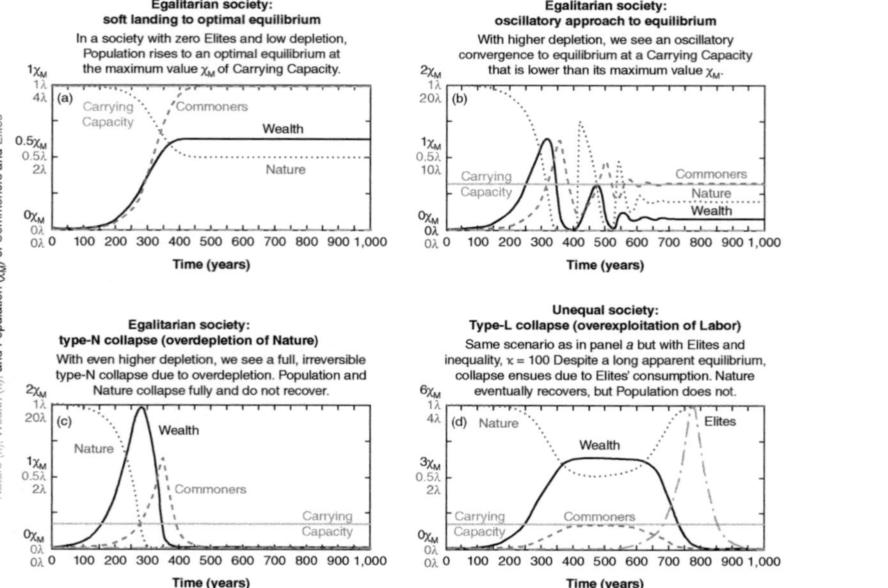

1. A dynamic approach to carry capacity analysis and various futures using the HANDY model. χ represents human population separated into Elites, χ E, and Commoners, χ C; y is Nature; and w is accumulated wealth. Nature, y, regenerates according to the logistic equation, with λ denoting nature's capacity. γ is the regeneration rate of nature, and δ is the depletion

- A sustainable steady state is shown to be possible in different types of societies.
- But over-exploitation of either labour or nature results in a societal collapse.

The model output shows not only that the inequality in the Human System increases the depletion of these planetary subsystems but also that the associated over-exploitation of common people is itself an independent threat to the future trajectory of the Human System. These two processes—over-exploitation of nature and overexploitation of labour—combine and exacerbate the threat of societal 'collapse'.

As the authors of this paper contend: 'Models have inherent inaccuracies and are beholden to myriad assumptions and simplifications but what they do show is that human civilizational collapse is possible, and that even a new resurgence would lead to a very different world in terms of human quality of life.' The quest for natural resource extraction to improve perceived well-being has been fundamental to the rise and fall of civilizations and remains a key feature of grand narratives around sustainability.

Easter Island: from tragedy to triumph of the commons?

Much myth and folklore surround the fate of the inhabitants of the triangular volcanic island in the South Pacific that had its first European visit on Easter Day in 1722. The Dutch explorer Jacob Roggeveen documented his 'discovery' of the island, where he spent a week and estimated there were 2,000 to 3,000 inhabitants. However, archaeologists suggest that the population may have been as high as 10,000 to 12,000 a few decades earlier. Roggeveen reported 'remarkable, tall, stone figures, a good 30 feet in height', fertile soil and a good climate, and 'all the country was under cultivation'. Subsequent fossil-pollen analysis suggests that the large trees on the island vanished by 1650, most likely cut down by the inhabitants for fuel or infrastructure. It is such research that has given rise to the popular narratives of 'societal collapse' caused by deforestation and ecological decline, and a subsequent crash in the population of the island.

Unfortunately, the 'first encounter' between the islanders and Europeans was also violent. Even in the first visit by the Dutch, a fight broke out with the locals within a week and several islanders were killed. Subsequent European visitations caused further harm to the population through diseases to which the islanders were not immune and the lamentable but usual tale of colonization. Yet these islanders, who call their land 'Rapa Nui', did survive the crises caused by their own ecological mismanagement, as well as the spoilage of colonial conquest.

I visited Easter Island some years ago after reading Jared Diamond's best-selling book *Collapse*, expecting a dystopian landscape of despair. Yet I was surprised to find that, contrary to the popular imagination, the indigenous inhabitants of Rapa Nui were thriving and still formed a majority of the island's population of 7,500. They had maintained their language and Polynesian cultural identity, though there was clearly inter-marriage with settler Chileans as well. Although the island's ecological diversity was now reduced to only around 40 species of flora and fauna, it was still a viable locale for human habitation. Long gone was the Rapa Nui palm—the world's largest by some estimates—and in its stead were many introduced species of plants. Sheep introduced by settlers had made their indelible mark on the island's surface, with pastures dominating the terrain. The occasional groves of trees were planted as wind breaks rather than for their existential value.

Although there was a massive disruption in the lifestyles of these early Polynesians, they were able to bounce back and to this day survive on the island. Environmental determinism around civilizational collapse needs greater clarity and needs to take into account adaptive features such as cultural changes in diet, habitation practices, and seasonal or permanent migrations. As recent research by anthropologists Robert DiNapoli, Carl Lipo, and Terry Hunt has shown, the inhabitants of Rapa Nui were able to adapt to deforestation that was prompted by natural changes in rainfall and temperature, as well as anthropogenic factors. They went through times of stress but bounced back as a mark of resilience within a socio-ecological system. Figure 2 shows the most recent evidence of how the islanders bounced back in

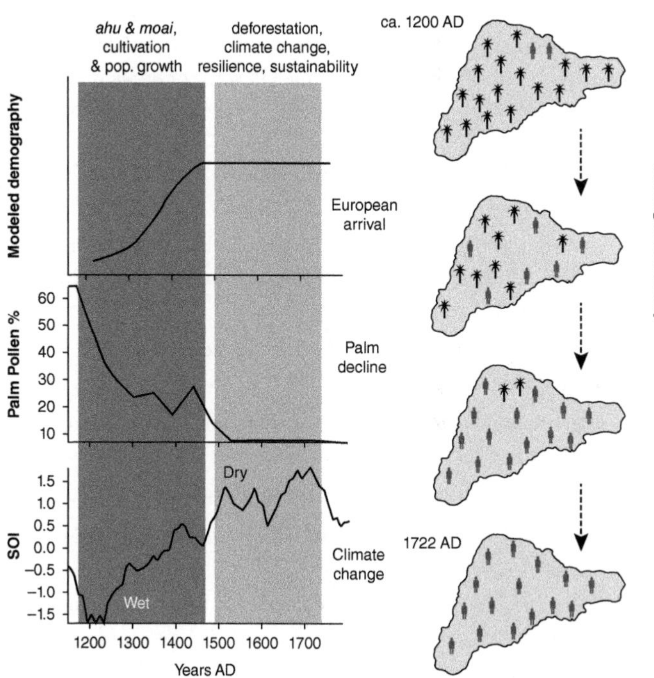

2. Human–environment interactions on Rapa Nui. Populations were stable, resilient, and sustainable despite their isolated location, marginal environment, and changing climatic and ecological conditions. SOI refers to the Southern Oscillation Index, which measures climatic variation in the South Pacific.

spite of ecological stress, contrary to the popularized but errant perspective of resource-linked demise of the population.

Rather than being a story of a tragedy of the commons, this research suggests that innovative small-scale communal practices were followed by the islanders to address the stresses on the ecological system. This approach is in synch with the Nobel Prize-winning work of Elinor Ostrom on governing the commons through cultural norms that avoid ecological collapse. Evidence of water conservation practices, mulch gardening of taro, and walled banana gardens as the climate presumably became drier suggests that these communities were able to follow Ostrom's eightfold design principles, which have been simplified in some of her later collaborative work with organizational psychologists as follows:

1. The group identity and its shared resource(s) should have clearly defined boundaries.
2. Individuals must not receive benefits from the group that are disproportionate to the contribution costs they have incurred; that is, high status or inequality must be earned.
3. Decisions must be made by consensus, collective-choice arrangements.
4. There must be monitoring by group members to deter free-riding and/or rule-breaking.
5. The severity of punishment for rule-breaking should be gradual.
6. There must be fair and collectively agreed-upon conflict resolution mechanisms.
7. Groups should be relatively autonomous and able to conduct their own affairs.
8. Larger-scale interaction requires appropriate coordination among sub-groups.

The parable of Easter Island shows us that sustainability is not a destination but a journey which civilizations take through a series of iterative learning processes. Nevertheless, what can happen on a small scale in Rapa Nui clearly cannot be replicated with ease at planetary scales. Far more intricate indicators and early warning systems are needed to ensure coordination of outcomes. There is also existing inequality of resources and historical processes of plunder from one society to another which have provided a much less 'level playing field' at the global level on which to govern the commons.

Ecological boundaries, social foundations, and sustainable development

Early efforts at global coordination of environmental policies for sustainable resource usage were targeted at very specific crises. The International Agreement for the Regulation of Whaling, which was signed in 1937 and is considered the first major multilateral environmental treaty, forms a good example. This agreement was further strengthened by the establishment of the International Whaling Commission and the formalization of the International Convention for the Regulation of Whaling in 1946. Despite the treaty having its ebbs and flows of national membership over the past several decades, the resilience and recovery of the world's largest mammals have been remarkable. A recent study of recovering whaling populations, published in the *Proceedings of the Royal Society, London*, showed that the humpback whale population in the southern oceans has gone from a low of 400 in the 1950s to over 25,000 in 2019.

However, such ad hoc crisis response agreements did not consider planetary processes and global environmental change until the advent of the 1972 Stockholm Conference on the Human Environment. Despite Cold War rivalries preventing consensus, with a boycott from both East and West Germany, the conference led to the establishment of the United Nations Environment Programme (UNEP). The General Assembly resolution which established UNEP did not use the word 'sustainability', nor was there any mention of ecological indicators, but rather a commitment to 'safeguard and enhance the environment for the present and future generations of man'.

Some of the consternation over the formation of UNEP came from developing countries who were concerned that poverty alleviation efforts would be constrained by concerns about environmental decline. Linking environmental conservation with poverty alleviation was what brought forth the emergence of the term 'sustainable development'. An earlier version of the concept was enshrined first in the World Commission on Environment and Development, chaired by erstwhile Norwegian Prime Minister Gro Harlem Brundtland, which published its landmark report in 1987. The need to consider a temporal aspect and limits to human activities motivated the shift to then using the word 'sustainable development'. However, initially there was a tendency to focus such efforts on developing countries through initiatives like the eight Millennium Development Goals (MDGs), which were set forth between 2000 and 2015.

When a stocktake was done of the MDGs in 2015 the results were mixed, leading to the realization that developed countries also needed to be held to the same standards in order to truly achieve planetary 'sustainability'. Figure 3 shows the evaluative results from key features of these goals as prepared by the independent non-profit organization The Center for Global Development. The original eight MDGs had 'environment' lumped as one category and, as a sign of the times, mitigation of the AIDS pandemic, especially in sub-Saharan Africa, was highlighted along with malaria as one of the goals dealing with disease mitigation. Since AIDS is now a largely manageable chronic disease, this focus has shifted and new challenges such as COVID have emerged. Indeed, the higher morbidity and mortality from COVID in developed countries also indicated that 'development' is not just a 'developing country' issue.

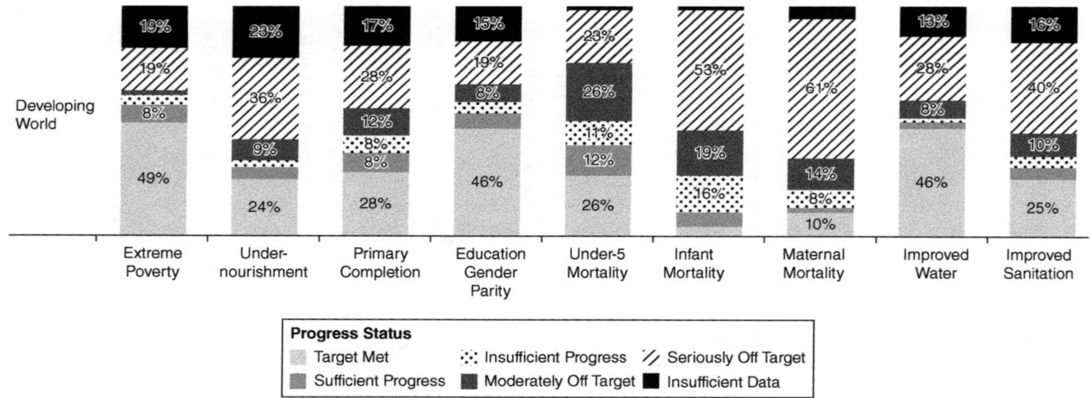

3. Evaluation of the Millennium Development Goals by category.

When a new generation of goals came to be developed in the next 15-year cycle, there was a recognition that all countries needed to be included in the development enterprise. In 2015, the United Nations thus put forward 17 'Sustainable Development Goals' (SDGs) with 147 targets to be met by 2030. These goals are now ubiquitous, and a wide range of monitoring and evaluation efforts are ongoing with the goals, which can be read online (see Further reading). A key feature of the SDGs was to promote a convergence of top-down and bottom-up approaches to meeting the targets. The Sustainable Development Solutions Network (SDSN) was also developed in parallel by one of the architects of the SDGs, Jeffrey Sachs. The SDSN comprised members from civil society, city governments, and private sector entities and was a 'big tent' that could work collaboratively on meeting targets at different scales.

Despite such inclusive efforts, there has been considerable tension in both academic and policy circles about the SDGs. The development economist William Easterly noted that SDGs should stand for 'Senseless, Dreamy and Garbled', while renegade social scientist Bjorn Lomborg asserted that all goals needed prioritization based on impact. By this measure, he suggested that trade liberalization and contraception access should be the highest-priority targets. His main ecological target which merited prioritization was reducing coral reef loss. This ecosystem trumped others such as tropical rainforests because the return on investment was greatest in comparison with use value.

Another means of prioritization was also set forth by natural scientists through the nomenclature of 'planetary boundaries'—popularized by the work of systems scientists Will Steffen at the Australian National University and Johan Rockstrom during his directorship of the Stockholm Resilience Centre. In a highly cited paper in 2011, Steffen and Rockstrom assembled 21 other scientists as coauthors to suggest that nine factors were particularly significant in maintaining sustainability of life as we know it today. The following are the key parameters they have set out so far and they continue to work on better quantification of some of these boundaries:

1. climate change (CO_2 concentration in the atmosphere < 350 ppm and/or a maximum change of +1 W/m^2 in radiative forcing);
2. ocean acidification (mean surface seawater saturation state with respect to aragonite $\geq$ 80% of pre-industrial levels);
3. stratospheric ozone depletion (less than 5 per cent reduction in total atmospheric O_3 from a pre-industrial level of 290 Dobson Units);
4. biogeochemical flows in the nitrogen (N) cycle (limit industrial and agricultural fixation of N_2 to 35 Tg N/yr) and phosphorus (P) cycle (annual P inflow to oceans not to exceed 10 times the natural background weathering of P);
5. global freshwater use (< 4,000 km^3/yr of consumptive use of runoff resources);
6. land system change (< 15% of the ice-free land surface under cropland);
7. the erosion of biosphere integrity (an annual rate of loss of biological diversity of < 10 extinctions per million species);
8. chemical pollution (introduction of novel entities in the environment—not yet quantified); and
9. atmospheric aerosol loading (not yet quantified).

The concept and its metrics are widely debated by scientists and gained traction with some notable global initiatives such as the Earth Commission. At the same time, concerns about minimum social foundations for human development and the right to have access to certain key resources such as water are just as compelling. The compromise zone has emerged in the form of 'Doughnut Economics', as popularized by economist Kate Raworth. She has developed a useful heuristic which combines planetary boundaries at the outside of the doughnut and social foundations within the

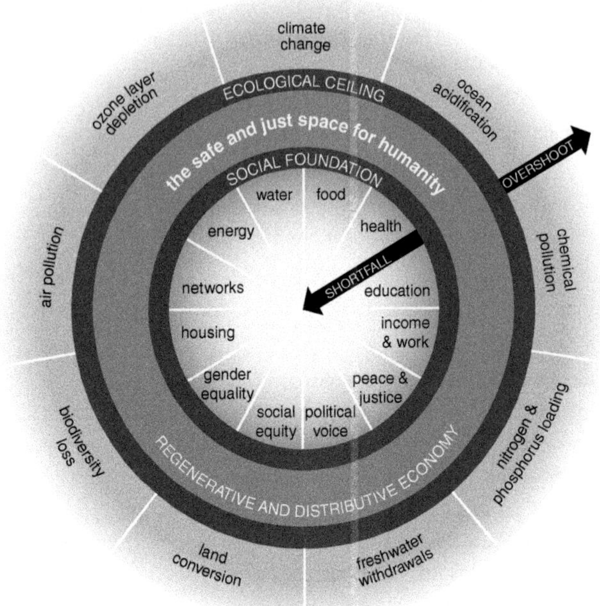

4. Doughnut Economics paradigm combining planetary boundaries and social foundations.

inner circle and posits that the corpus of the doughnut itself is the 'safe and just space for humanity' (Figure 4). This model has been embraced by numerous European jurisdictions as well as by cities such as Amsterdam to set their own targets for development, and also has appeal for sustainability education.

Either end of the political spectrum continues to criticize these attempts at operationalizing sustainability from a natural science and social science perspective. Those on the political Left consider the approach to be naive in terms of entrenched power dynamics around capitalism. The Netherlands Environmental Assessment Agency, a research institute that advises the Dutch government, sees a €7 billion opportunity for the Dutch economy in the transition to Doughnut Economics. However, there is concern that the process would stall and eventually be co-opted by rent-seeking interests. Tiers Bakker, a councillor with the Dutch Socialist Party, remarked: 'We don't need doughnut economics. We need bagel socialism!' On the other end of the political spectrum the concept has been presented as limiting economic growth and enterprise because of constraints set forth by boundaries which are at times not even measurable.

Using the 'Doughnut Framework', a team of researchers led by Andrew Fanning at the University of Leeds analysed the historical dynamics of 11 social indicators and 6 biophysical indicators across more than 140 countries from 1992 to 2015. They found that countries tend to overshoot biophysical boundaries faster than they achieve social thresholds. The number of countries for which the planetary boundaries were exceeded over this period went from 32–55 per cent to 50–66 per cent, depending on the indicator. At the same time, the number of countries achieving social thresholds increased for five social indicators (in particular life expectancy and educational enrolment), decreased for two indicators (social support and equality), and showed little change for the remaining four indicators.

At their core, the vigorous debates in this arena come down to whether conversion of natural capital to other forms of anthropogenic capital should be considered 'sustainable'. Natural capital would include fuel and non-fuel minerals, biotic resource for human consumption such as crops as well as environmental and ecosystem capital for sustaining life. Anthropogenic capital would include manufactured capital, human knowledge, education and social capital. Capital in both forms is an *asset* and services which they provide is a *flow*. This distinction is often misunderstood and has consequences in more detailed analysis of sustainability. Also worth noting is that financial assets are merely a corollary for measuring more fundamental stocks such as infrastructure or an educated populace. Furthermore, since sustainability is an emergent confluence of social and natural systems, it should be coupled with integrative terms and not reduced to parts. One of my mentors in this arena, William Clark, cautioned me hence that to be true to this emergent complexity we may say 'sustainable development / futures' but not 'sustainable finance' or 'economic sustainability'.

Scientists have struggled with having two categories of sustainability—weak and strong—to differentiate processes which are naturally sustainable (and hence strong) and those which may convert natural capital to service sector economies (weak). The latter can subsequently be sustainable on human timescales but will inevitably require some natural capital exchange with other sub-systems on the planet. The key to unpacking this differentiation is to consider the connections between material and energy flows on the planet, which will be explored in the next chapter.

Chapter 2
How energy and materials flow through systems

The concept of sustainability has currency based on what humanity values in the living world. We talk about 'sustaining life' on Earth—for that is what makes our planet special in the observable universe. We seek to find exoplanets that orbit at the right distance and have the right geology to enable life to evolve. Earth has been able to achieve life as a key 'emergent property' (an outcome with properties that are greater than the sum of their parts) because the planetary environment has been in a 'Goldilocks zone' to allow for matter to be 'energized'. Indeed, the attribute of life that is most palpable is an organized and coherent nexus between matter and energy. A living cell is a system nested within an organism; which is nested within an ecosystem; which is nested within a planetary system.

Planetary processes operate as a series of nested systems that are fundamentally predicated on energy and material flows. Sustainability requires us to consider how materials and energy are internally cycled through these systems and exchanged between them. When the Nobel laureate Richard Smalley, one of the discoverers of the 'buckyball' form of carbon, buckminsterfullerene, was asked to prioritize the world's top challenges for the next 50 years around the turn of the millennium, he noted energy at number 1. He reasoned that other issues such as poverty alleviation or even disease eradication require energy flows. But the amount of energy we need physically for metabolic processes is vastly less than what we now use for our social existence. We have built technologies and infrastructure to augment our quality of life and harnessed the most versatile manifestation of energy—electricity—to literally and figuratively 'enlighten' the planet.

Energy flows to the Earth are today ultimately traceable to the Sun, while energy stored within the Earth from its formation also fuels geological processes such as plate tectonics. Natural radioactive decay of elements in the Earth also provides some additional 'internal energy' to the planet. To understand sustainability as a concept we need to recognize that there are hierarchies as well as interdependencies between the material and energy flows within these subsystems. Unfortunately, the constraints and hierarchies of the material–energy nexus are often trumped by geopolitical considerations. This nexus refers to the intricate and interconnected relationship between materials and energy in various industrial, economic, and environmental processes. It underscores the mutual dependence and influence between the use of materials and the consumption of energy in these processes. The concept is crucial for understanding resource efficiency, sustainability, and the environmental impact of human activities. Countries aspire to be 'energy independent' or 'resource secure' within borders that are an artefact of history rather than the physical geography of nature. Furthermore, the sources of energy which we have predicted to come online and be effective in delivery have not followed their predicted path.

In the 1970s, a researcher named Cesare Marchetti started to consider different forms of energy that could be developed and suggested a 'substitution model' for transition from one form to another. Starting from wood as our earliest material for energy production, he considered coal, followed by oil, natural gas, and finally hydrogen as the ultimate fuel which could deliver. However, his predictions of the rise and fall of the various sources have up to now been far from accurate. The challenge remains one of finding efficiency in terms of both the material and energy input required to deliver the requisite amount of output.

Fossil fuels have exhibited remarkable staying power because of their efficiency in this regard. Another metric that is important here is *power density*—how much energy can be derived from a unit of material per unit time. Nuclear power comes out very favourably in this regard but there are numerous other social factors of risk perception which then render it less appealing to the public. Energy and material security is also defined in geopolitical terms rather than in terms of where it is most ecologically or even economically efficient to harness the resource.

Sustainable energy and rebound effects

During the height of the Cold War, when the military industrial complex was in full force, there was a remarkable recognition of global environmental vulnerabilities and the need to understand material and energy flows. After six years of negotiations, US President Lyndon Johnson and USSR premier Alexey Kosygin joined 10 countries to create the International Institute for Applied Systems Analysis in 1972. While the United Nations was focused on multilateral political mobilization through the creation of the UN Environment Programme, IIASA was focused on scientific mobilization. Austria was a neutral country in the Cold War and a stately Schloss outside Vienna in the town of Laxenburg was chosen as the headquarters. The establishment of IIASA opened the door for sustainability science research at a systems level, whereby the material–energy nexus could be more effectively researched. Cesare Marchetti would also be employed at IIASA and further refined his substitution model there.

In 2012, IIASA completed the Global Energy Assessment (GEA), the first integrated assessment of world energy resources as well as a primer on challenges and opportunities for harnessing the material–energy nexus towards a more sustainable future. The GEA database includes detailed quantitative information for 41 pathways to reaching sustainability, of which six pathways meet the goal of providing 'sustainable energy for all of our planet's population' (SEforALL). There are three objectives: providing universal access to modern energy services; doubling the global rate of improvement in energy efficiency; and doubling the share of renewable energy in the global energy mix.

While the first law of thermodynamics, in its most simplistic formulation, suggests that 'energy can neither be created nor destroyed', we have functionally 'created' means of harnessing energy from matter. The combustion of fossil fuels, the splitting of uranium atoms in nuclear fission, or the absorption of solar heat by specialized materials, are all examples of such ingenuity. A planetary vision of sustainability thus requires us to consider the material–energy nexus in fundamental ways. Basic thermodynamic principles are essential to understanding sustainability.

On a planetary scale, the Sun is our ultimate source of energy, but there is also geophysical energy stored within the Earth's interior. Another key concept in understanding the material–energy nexus is *entropy*, which defines the distribution of energy in a system in terms of its availability for useful work. The less available the energy is, due to the extent and quality of order in a system, the higher the entropy. Photosynthesis has been a key feature in channelling solar energy towards life. However, humanity has had a major intervention in this process; and by one estimate nearly 40 per cent of potential terrestrial net primary productivity (rate of energy availability from plants) is used directly, co-opted, or forgone because of human activities.

All these various mechanisms of providing consumers with more useful information to assist in decisions which are more societally efficient bring forth another important concept for our consideration in defining optimal economic order. The concept of 'exergy', which is essentially the useful, available energy to undertake work, is helpful in refining economic equilibrium analysis to better accommodate sustainability conditions. Although it was not explicitly mentioned at the time, the great ecologist Howard Odum presented energy and material flows through ecosystems as a means of efficient resource usage. At its core, exergy gives us the maximum theoretical work obtainable from an overall system, considering that the system will come into equilibrium with its environment, at which point no further work can be obtained from it. There are three different origins of exergy flows on Earth: the energy flow from the Sun that is utilized by life and movements of air and water through the planet; the energy flows from radioactive decay within the lithosphere (the Earth's crust and upper mantle); and the tidal phenomena caused by gravity and rotations within the solar system. However, the solar energy flows dominate by an order of magnitude of several thousand over the other two sources.

In some ways the Earth can be thought of as a giant fuel cell to convey the linkages between energy, entropy, and material flows. Some parts of Earth are also storing energy like a battery but a fuel cell metaphor is more appropriate as there is dynamic flow and conversion of resources. Unlike batteries, fuel cells require a source of energy for continuous input of fuel—which in the case of the Earth is sunlight or geothermal energy.

On the surface of the Earth, there are particular points at which energy from the planet's interior is released (such as hydrothermal or alkaline vents), creating opportunities for the emergence of life. The gas hydrogen, the lightest of the elements, is lost from Earth's atmosphere due to solar and ionic neutralization processes and creates an 'energy channel'. Harnessing the 'useful work' out of this channel

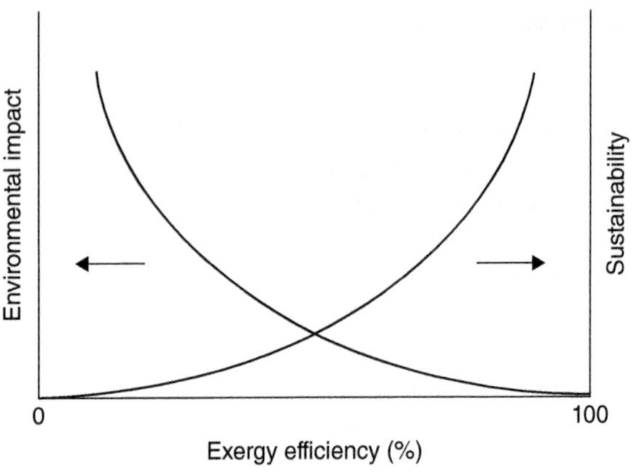

5. Relationship between exergetic efficiency, environmental impact, and sustainability.

provides us with what is termed 'exergetic efficiency'. Figure 5 lays out the relationship between environmental impact and sustainability in the context of such a metric, which is also referred to as 'rational efficiency'.

Much of the growth in human technological development and population that happened during the three major industrial revolutions can be seen through the lens of this quest for exergetic efficiency. The first industrial revolution focused on moving from manual and animal labour to the invention of the steam engine (early to mid-19th century). The second industrial revolution towards the end of the 19th century was brought about through large-scale electricity production, which led to the development of myriad technologies. The third industrial revolution, from the 1950s to the 1980s, further enhanced the speed and efficiency of utilizing electric energy for information processing through the development of computational technologies. The use of novel materials such as silicon semiconductors was crucial in furthering the exergetic efficiency of this phase. On a per-unit basis environmental impact has clearly declined and we have increased the potential for a more sustainable outcome in congruence with Figure 5.

However, per unit efficiency can be a deceptive measure if we do not consider the overall aggregate consumption change which may come from more efficient technologies. For example, hybrid cars are efficient in terms of fuel mileage but can potentially lead to drivers feeling more sanguine about their use, and driving more than they might otherwise. This 'rebound effect' was noticed as early as the 19th century by the coal economist William Stanley Jevons in the context of efficiency improvements in steam engines (hence it is also known as the 'Jevons paradox').

One of the core reasons for our lack of realization of this paradox is that the price for using resources might not be enough to trigger behavioural change for a variety of reasons. Economists call the sensitivity of consumption behaviour to price 'elasticity'. In rural areas you may have little choice but to drive cars even if the price of fuel goes up and hence the system would be 'inelastic'. Efficient systems can reduce consumption if we keep track of aggregate demand and try to ensure that we can find a pathway for reducing consumption. Residents of rural areas who own a hybrid car would only be more efficient if they use less fuel in total than those who do not have such a car. We also need to create price incentives for country dwellers to consider ride sharing or other means of mitigating their total resource imprint.

Furthermore, the 'impact' axis needs to be further unpacked, as some environmental impact mitigation strategies can lead to other impacts. Currently, there is a consensus around carbon emissions as being a key impact variable which is a 'threat multiplier' for human well-being due to climate change. Hence many scenarios around energy delivery transitions to avoid the rebound effect are focusing on this variable in terms of the impact of emissions on global temperature changes.

Carbon mitigation as a touchstone of sustainability

The International Energy Agency's '2DS' scenario, which was prepared in 2017 as part of their Energy Technology Perspectives report, takes into account a 70 per cent reduction of CO_2 emissions in the energy sector from today's levels by 2060. It is a highly ambitious scenario where there is a 50 per cent chance of limiting the temperature increase by 2100 to only 2 °C. Figure 6 lays out how various exergy flows would most likely need to play out from fuels to economic output or infrastructure. There are still sizeable conversion losses in this scenario, which highlights the challenge posed by the second law of thermodynamics. In one of its forms this law suggests that 100 per cent efficiency is impossible at temperatures above absolute zero (–273.15 °C, when molecular motion ceases).

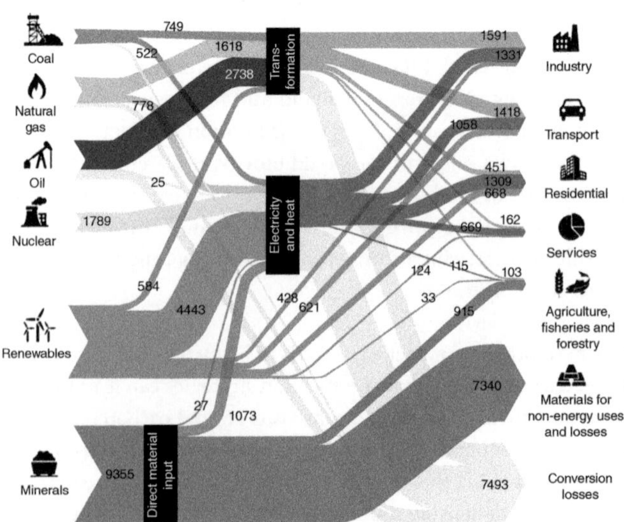

6. World exergy flow analysis for the IEA 2DS scenario for 2050. All data are expressed in million tons of oil equivalent (MTOE).

From the perspective of sustainability science, we should also consider the operational efficiency of materials such as semiconductors for quantum computing and other key technological applications. How much energy is required to cool certain materials to improve their efficiency needs to be considered at a systems level. The same is true of nuclear fusion research for energy generation, where the reactions—thus far—need more energy injected than the return. Finding magnets and superconductors which can deliver nuclear fusion reactions at commercialized scale with less energy injection than output is the ultimate 'holy grail' of limitless clean electricity generation. The same simple input–output measurements across the wide swathe of technological solutions are core to upscaling solutions which may be deemed 'sustainable'. Input–output measurements as a means of measuring productivity can also be translated in economic terms and were at the heart of the work of Nobel laureate Wassily Leontief. Among his many qualities was the willingness to collaborate with colleagues across disciplines. His mentoring of Faye Duchin, a young computer scientist with interests in ecological flows, led to a deeper connectivity between the economics of technological change and work on sustainability.

The French-American systems scientist Robert Ayres is credited with giving quantitative rigour to linking material flows to economic analysis within the eclectic realm of sustainability science. Ayres started his career as a physicist at the University of Chicago and ended up retiring as professor of environmental management at INSEAD (a notable French business school). In 2005, he was commissioned by IIASA to estimate the 'Mass, Exergy, Efficiency' in the current US Economy. His analysis was subsequently peer reviewed and published and remains the only such analysis (Figure 7).

Ayres noted in his analysis that there are five types of work: namely muscle work by humans or animals, mechanical work by stationary or mobile heat engines (prime movers), and heat,

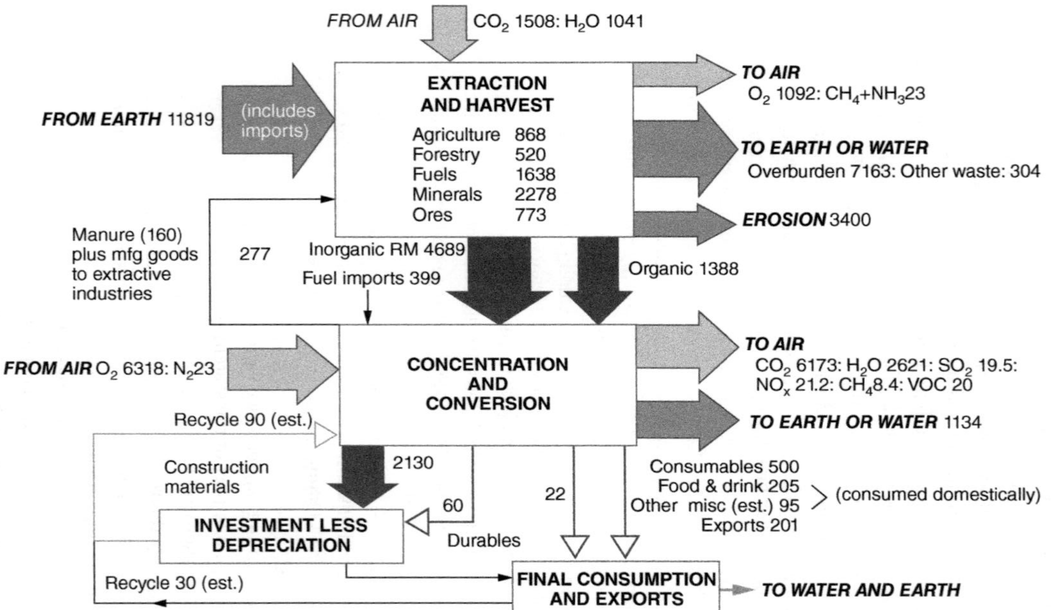
FROM AIR
CO2 1508: H2O 1041
EXTRACTION AND HARVEST
Agriculture 868
Forestry 520
Fuels 1638
Minerals 2278
Ores 773
FROM EARTH 11819
(includes imports)
TO AIR
O2 1092: CH4+NH3 23
TO EARTH OR WATER
Overburden 7163: Other waste: 304
EROSION 3400
Manure (160) plus mfg goods to extractive industries
277
Inorganic RM 4689
Fuel imports 399
Organic 1388
CONCENTRATION AND CONVERSION
FROM AIR O2 6318: N2 23
TO AIR
CO2 6173: H2O 2621: SO2 19.5: NOx 21.2: CH4 8.4: VOC 20
TO EARTH OR WATER 1134
Recycle 90 (est.)
Construction materials
2130
60
22
Durables
Consumables 500
Food & drink 205
Other misc (est.) 95
Exports 201
(consumed domestically)
INVESTMENT LESS DEPRECIATION
Recycle 30 (est.)
FINAL CONSUMPTION AND EXPORTS
TO WATER AND EARTH

either at high temperatures (for metallurgical or chemical processes) or at low temperatures for space heating, water heating, etc. The ratio of output work to input exergy is the thermodynamic efficiency of the conversion process. Tracking such inputs and outputs over time is a daunting task and a snapshot approach as presented here from data that is a few decades old is likely to not be very useful for sustainability planning purposes. However, if we can identify the key nodes of measurement and invest in monitoring technologies to provide real-time data we could be closer to effective management of the complex adaptive systems of our planet.

Complex systems: scale, time, and sustainability

When we say a system is 'complex' it has very specific scientific features which differentiate it from being 'complicated'. The easiest metaphor that is often used to differentiate complicated from complex systems is that while both may have many parts, complicated systems have additive predictability whereas complex systems have unpredictable 'emergent' properties. A car engine is complicated, with many parts, but once you put each part together, it has additive functionality. If one part of the engine stopped working, we could go in and fix that part and it would work again.

With complex systems, we cannot predict what would happen exactly if one part of the system stopped working. Further, even though there are fundamental mechanical equations which govern physical properties of the system, slight inevitable variations lead to highly divergent outcomes. Such a property is referred to as 'sensitive dependence on initial conditions'. These variations can, however, smooth out in long-term trends of the macro-level behaviour of the system. This is why we often cannot predict the weather with accuracy beyond 10 days, but can predict long-term climate trends with greater confidence.

If the system is adaptive, it could also have some process of repair which could lead to other states that would also be predictable. Mammalian immunity is a sterling example of an 'adaptive' system that can learn to repair in response to shocks. Planetary processes are 'complex' because they are not additive in their outcomes, so gauging measures of sustainability requires us to consider multiple possible pathways. Complex systems do have certain 'attractors'—which are zones of some convergent stability. Although inherently 'chaotic' and unstable, complex systems usually evolve towards a number of steady states which constitute these 'attractor' basins. The steady state or equilibrium achieved in these attractor basins is usually achieved in natural systems through various material flow cycles that involve an interaction of biological, chemical, and geological properties.

We usually refer to these 'biogeochemical cycles' of the different elements as separate 'cycles', such as the carbon, nitrogen, phosphorus, etc. cycle. However, in reality all these cycles are connected and lead us to a particular steady state in a natural system. On planetary scales, these cycles operate physically between land, air, and water. Figure 8 shows the metaphorical representation of these cycles as a mechanical turbine system between these three different phases. However, in the case of the biogeochemical cycle, the 'turbine blades' are driven by a biologically determined 'shaft'. The reverse applies in the case of the steam turbine, where natural processes such as wind or water turn the turbine, which moves the shaft.

Going back to the differentiation between complicated and complex systems, the system shown in Figure 8 also has another important feature which can lead to emergent properties—the notion of coupling between the various phases. The mechanical analogy gives a simple way of illustrating that the two primary biogeochemical cycles, that of the terrestrial and oceanic systems, although driven largely independently, are 'coupled' through the atmosphere. This coupling is analogous to the connecting belts

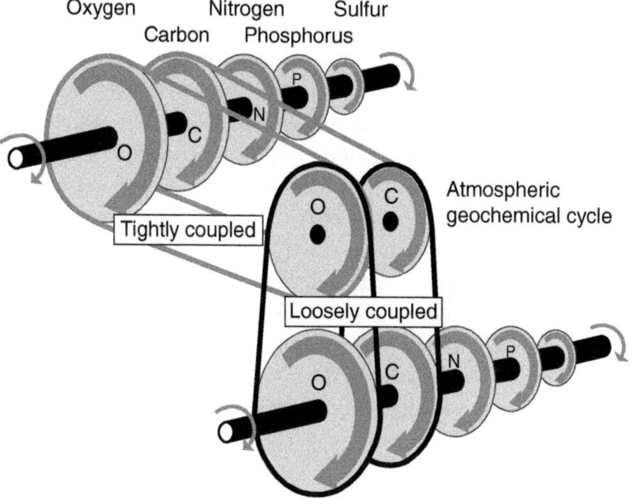

8. Connections between biogeochemical cycles.

between the turbine gears being tight or loose. In a purely mechanical system this could be fixed, but in a complex system this coupling can be more dynamic and emergent. The atmospheric cycles (primarily of oxygen and carbon dioxide) might be considered like idler cogs in a gear train. Therefore, even though the connection between the terrestrial and oceanic biogeochemical cycles mediated by the atmosphere is loose, that of oxygen is consequential on timescales of millennia. The islands of stability over geological time get created through such emergent properties of the interaction of these cycles.

Through geological time, there have been many steady states of biogeochemical cycles which have been less receptive to human life. So even though there may be an equilibrium in that planetary state it is by no means a 'sustainable' outcome from the perspective of humanity. For example, an arid, desert environment may be a steady state outcome of planetary processes but it is not as receptive to human life. We are often tempted to consider the term 'equilibrium' as somehow synonymous with 'sustainability'. Both terms connote a steady state, but equilibria at the biochemical level can prevent adaptive mechanisms towards sustainability. When we move from one 'attractor basin' to another, it is also likely to be an irreversible change. Indeed, research on historical ecological shifts from one attractor basin to another by the late Canadian systems ecologist James J. Kay in the 1990s could not find a single example of any ecosystem returning to its original state following a disruption. Even when we have geological cycles such as ice ages and interglacial periods, the ecosystems which they house do not return to original states at the end of a cycle—indeed that has been one of the key drivers of extinction and speciation phenomena over geological time.

Another key feature of natural systems in this regard is how they are scaled, and the disjuncture that can happen between growth patterns of natural systems and social systems that humans may initiate. Physicist Geoffrey West, who served as the president of the Santa Fe Institute (the world's leading think tank on Complex Systems), has noted that among the key reasons human beings have had difficulty in arriving at a 'grand unified theory of sustainability' is the challenge of different rates of scaling between natural and social systems. In his landmark book *Scale*, West notes that while biological systems follow 'sublinear scaling' in their growth trajectory (e.g. as organisms increase in size, their metabolic rate tends to increase at a slower rate than would be expected based on a simple linear scaling), social and economic systems can follow a 'superlinear scaling' (e.g. increasing return to scales in corporations or in cities' growth patterns). Intriguingly the scaling difference for social and economic systems is always around 15 per cent above the expected linear equivalent and is related to underlying structural properties of complex systems.

This leads to a fundamental disjuncture in planning for sustainability because mathematically superlinear systems have what is termed a *finite time singularity*—which is in some ways similar to a phase transition in physics. The growth curve in this situation is not like an exponential population growth curve but far more acute in its growth, such that within human timescales it approaches infinity. Since such an outcome cannot be sustained with growth rates in biological systems, a sudden collapse becomes inevitable unless there is an intervention.

Such an intervention would involve either a major technological breakthrough or some other way of mitigating the growth pattern through reduced consumption. Figure 9 shows how Geoffrey West envisaged a pathway out of such a situation.

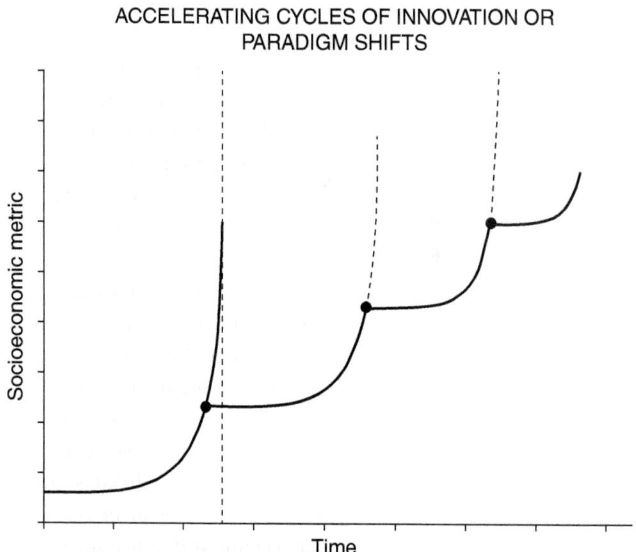

9. Technological ways of preventing a finite time singularity leading to a concomitant biological collapse. The black points indicate innovations which reset the curve.

Of course, the other way of preventing such an eventuality would be to simply reduce the growth rate of the socioeconomic metric. West goes on to state this as the following theorem: 'sustained open-ended growth in the light of resource limitation requires continuous cycles of paradigm-shifting innovation'. Furthermore, the more challenging aspect of the win-win trajectory is that the time between successive innovations also has to get shorter and shorter. Discoveries, adaptations, and technological change must occur at an accelerated pace for such growth to be sustainable.

The next question to ask is whether the social and economic pathways being followed by human societies are actually following a superlinear pathway. Recent research conducted by astrophysicist Anders Johanssen and complex systems scientist Didier Fornette suggests that even with reduction in growth of population, the economic growth demands are indeed following a superlinear pathway. By their calculations we would reach a finite time singularity by 2052, at the same critical time for population and economic growth levels (with an error margin of 10 years). A radical innovation shift would thus be needed by then to prevent a dramatic spiral towards unsustainable outcomes. West uses the metaphor of a treadmill:

> We are not only living on an accelerating treadmill that's always getting faster and faster, but at some stage we have to jump from one onto another one that's going even faster. And this entire process has to be repeated into the future at a faster and faster rate.

While there is no way to predict the pace of innovations, futurists such as Ray Kurzweil have tried to track past innovations and project key technological leaps which may yet occur in our future. Human ingenuity has led to a remarkably accelerated pace of innovations because we have harnessed some key structural properties of matter and energy to allow for this 'Great Acceleration' of technological development. The combustion of matter to get mechanical kinetic energy through steam fuelled the industrial revolution. This was followed by electricity delivery, harnessing the wonders of the electromagnetic force. This in turn allowed for the computational revolution. Kurzweil and other futurists have taken the mathematical usage of the term 'singularity' and suggested the ultimate way forward for bountiful innovation to allow for endless economic growth would be the rise of artificial intelligence and related infrastructure such as quantum computing. In such a scenario human biology would be augmented by genetic alterations and nanotechnology that would allow artificial intelligence systems to essentially make us cyborgs. In a dystopian scenario this could lead to a collective intelligence like 'The Borg' in *Star Trek*, but in a utopian scenario it could lead to a world where technology is able to calibrate our existence towards sustainability.

For transitions to be sustainable in democratic societies, there will need to be a focus on justice for communities that have faced historic deprivation. In the short term this may mean that there will be some environmental impacts of equalizing development outcomes. But in the long term, such a 'just transition' is likely to be more sustainable because it would meet ethical obligations to communities and build trust across society, mitigating the potential for violent social conflicts. While some social conflict can be important for positive transformation of societies, violent and destructive conflict that ensues from a lack of trust can undermine sustainability by leading to securitized states. Resource usage in such securitized environments of war has been among the most serious threats to human sustainability in the last century, particularly during the Cold War and the nuclear arms race.

Considering past violent conflicts during times of industrial change, the European Union has established a 'Just Transition Mechanism' that links the 'Green New Deal' with livelihood development, reducing energy poverty and providing affordable and efficient housing. The United Nations Development

Programme has also noted the importance of such linkages. The renewable energy sector has given considerable hope for such a 'just transition' in the developing world, where several hundred million people still need access to electricity. In 2019, employment in renewable energy was estimated at 11.5 million worldwide, with women holding 1 in 3 of these jobs. And the numbers are growing, with the International Renewable Energy Agency projecting that renewables could employ more than 40 million people by 2050. Human societies have developed norms of behaviour to manage technological transitions. Incentives that spur innovations are linked to what we have come to define broadly as economic processes. As we have developed new technologies our economic system and the supply chains that link primary natural resources to goods and services provision have also become more complex. The term 'technoeconomic analysis' is used to consider such analyses, and we will unpack this in the next chapter.

Chapter 3
Technological and economic interventions for a sustainable society

The allure of technology as a solution to our sustainability challenges remains highly dependent on social behaviour. By one estimate made by Geoffrey West, an average human needs around 90 watts of power for biological sustenance but around 11,000 watts for 'social well-being' in even an efficiently managed developed economy. Social well-being encompasses all the variety of amenities we have grown to consider important for connecting with each other, such as electronic devices, as well as key services that make us feel civilized, such as a haircut or an education. Therefore, humans require far more energy and materials for their social existence than their metabolic needs as an organism. However, the social and the biological are intertwined in terms of our survival as a species. This is exemplified by high rates of suicide in countries which meet the biological needs of their citizens but may be deficient on social capital. Technology is a key mediating mechanism to managing human needs for well-being in both biological and social terms, but it can itself have an ecological impact as well on planetary support systems.

Engineers are often at the forefront of positing technological solutions but there are also many social scientists engaged in this space and a vibrant academic field of 'Science and Technology Studies' (STS) has developed over the past several years to consider the social implications of technological interventions. At the same time 'technoeconomic analyses' are also carried out to weigh costs and benefits of particular interventions. Within the context of sustainability research, these scholars consider the full spectrum from scepticism to optimism about technological solutions.

The 'techno-optimists' have come under considerable criticism from social scientists working on sustainability in two primary ways. First, there is a concern for access to technology in a world which has such stark levels of economic inequality. Second, there are concerns about the social impacts of the technology itself and whether this impact can be determined effectively in advance of implementation. However, the techno-optimists have also come under criticism from innovators in engineering design. One of the pioneers of this field, John Ehrenfeld, who had burnished his credentials as a systems engineer at MIT, became a major critic of the techno-optimists in his book *Sustainability by Design*. He declared that a culture of sustainability should have 'techno-skepticism' as a trait alongside being biocentric and communitarian.

Ehrenfeld's influential book used causal loop diagrams developed by his colleague at the Sloan School of Management, Peter Senge, to consider the perils of relying on technology to achieve sustainability goals. Ehrenfeld offered a new definition of sustainability: 'the possibility that human and other life will flourish on the planet forever'. Perhaps if we consider extraplanetary travel the 'flourishing' could also be stretched over wider timescales. He further claimed that 'almost everything being done in the name of sustainable development addresses and attempts to reduce unsustainability. But reducing unsustainability, although critical, does not and will not create sustainability.' Ehrenfeld goes on to use the Zen metaphor of the glass half-full and half-empty as an example of a complementary spectrum that is often misleading when considering sustainability. Unlike the glass metaphor, reducing 'unsustainability' will not create space for

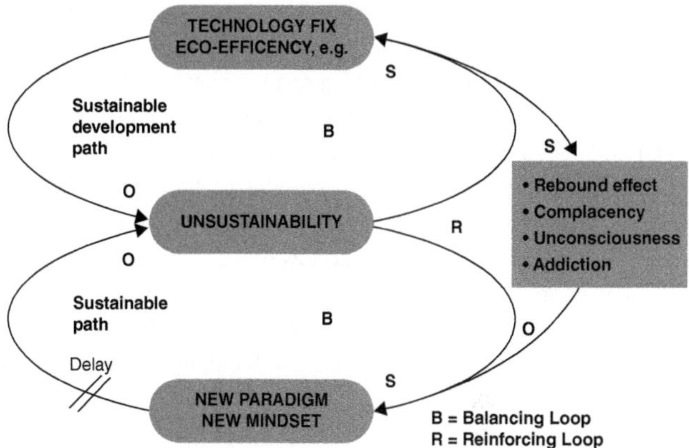

10. Ehrenfeld's causal loop critique of technology. The term 'unconsciousness' implies lack of awareness of one's actions.

'sustainability'. Emergent systems have this property that we cannot solve problems at the same level of thinking at which we created them.

Figure 10 shows Ehrenfeld's concern with the way in which technology is invoked as the solution in terms of a causal loop diagram. Sustainability is considered an emergent property that is an outcome of our choices which could have various manifestations. Technology can give us durability or reliability but not inherent sustainability. 'S' refers to trajectories for solutions that are often presented in terms of technologies and 'O' refers to opposing trends that are a by-product of those solutions. So, as the diagram shows, to get out of the negative spiral of the technology loop at the top, we need to consider the underlying symptoms of complacency, addition, rebound, and lack of conscious realization of our environmental impact. This would require a paradigm shift towards a more self-conscious systemic change in our relationship with natural resources, recognizing them as finite, and hence there are limits to what technologies can also attain.

For example, moving towards the technological fix like an electric car can still lead us down towards rebound and complacency. It can also lead to a higher level of aggregate resource usage which may come across as 'sustainable development' in the short term, as social and economic 'green economy' indicators show. However, ultimately, this too would be unsustainable on longer timescales. We can correct course from technological fixes by recognizing the pitfalls of pathway 'S' and head towards a new paradigm that could have a more sustainable outcome. However, there is a delay factor in terms of uptake of such a path and if we are not mindful of time and urgency we could still end up in an unsustainable outcome.

Since human societies use trade as a fundamental means of social interaction, we are presented with markets as an inevitable forum for wrestling with some of the more challenging aspects of sustainability. In their simplest forms markets are a means of connecting supply and demand for goods and services. William Rees, who has studied sustainability as a planner, suggests that technological solutions provide an easy supply-side framing of global environmental crises. Instead, if we focused on human behaviour as the core crisis, we would have a far more productive demand-side solution for sustainability challenges. Technology could still come to our assistance, but it would be targeted at changing human behaviour rather than at providing short-term 'solutions' that would lead us back into Ehrenfeld's unsustainability loop.

Technology and biology operate in different mechanistic spaces but if we are to follow the trajectory posited by Ray Kurzweil described in the previous chapter, technology could be implanted within the brain itself to modulate behaviour. While there would be serious security concerns about the misuse of such a technology by governments, such an approach could present us with the potential for positive social change through technological intervention. The challenge that lies before us is to choose those technologies whose use will lead to such outcomes. A next step to consider would also be how specific technologies not only mitigate human impacts but make the planet more viable for the diversity of non-human life as well. Might we thus consider a situation where the human footprint of technology is actually a positive systems contribution? Developing a means of assessing technologies for such potential must be a key next step in our quest for sustainability.

Technological assessment for a sustainable planet

Considering the rapid rise of technological tools during 'The Great Acceleration', there is a need to assess the quality of specific innovations in terms of their impact on the long-term viability of the planet. This harks back to the 'T' variable in the IPAT equation (discussed in Chapter 1) having a certain ambivalence which needs to be reconciled. To undertake this task the United Nations established an 'International Environmental Technology Centre' (IETC) in Osaka, Japan, in 1992, following a directive from the Rio Earth Summit. Over the past several years this centre has developed a methodology to assess technologies in terms of their sustainability outcomes.

There are six features of the IETC methodology for assessing technological sustainability:

1. addressing strategic as well as operational levels;
2. addressing sustainability (integration of environmental soundness, social/cultural acceptability, and technical and economic feasibility) through a specially designed methodology and criteria;
3. employing a progressive assessment procedure, through tiers addressing screening, scoping, and detailed assessment, thereby allowing entry points for a diversity of stakeholders and optimizing information requirements;
4. employing quantitative procedures that allow more objective assessment, sensitivity analyses, and incorporation of scenarios;

5. ensuring application to technology 'systems' as opposed to individual technologies; and
6. placing importance on information expertise and stakeholder participation.

To understand this methodology, let us use the example of synthetic biology as a genre of innovations that are currently gaining wide acceptance in sustainability circles. Synthetic biology (Synbio) may be defined as the construction of biological components, such as enzymes and cells, or biochemical functions or even full organisms that don't exist in nature. The process could involve a redesign of an existing organism's genetic structure through gene editing. This approach is different from the selective breeding of naturally mutated forms of a species or physical grafting of cultivars that humans have practised for millennia.

Regarding point 1 of the assessment framework given above, synthetic biology has a strategic advantage of upscaling at an accelerated pace, but this can also be considered a risk if the intervention goes awry. At an operational level, Synbio is gaining traction with the use of CRISPR gene-editing techniques that won the discoverers, Jennifer Doudna and Emmanuel Charpentier, a Nobel Prize in chemistry in 2019. The high level of control that is possible with this technique makes it operationally less risky. Artificial intelligence techniques can also be combined with Synbio approaches to craft particular enzymes that meet certain characteristics needed to deliver a particular outcome.

An integrated cost–benefit analysis which accounts for the full range of possible contingencies is then performed in step 2. This feeds into the stakeholder engagement process, by which there is transparency of social communication of the risk–reward trade-offs with community members. At this point it is important to monitor the potential for what geographer Roger Kasperson termed the 'social amplification of risk', which comes from any uncertain process in human perceptions. Effective communication and control of disinformation in the age of social media remains a key challenge in this regard.

Point 4 refers to the use of techniques such as Life Cycle Analysis (LCA) that can effectively evaluate key comparative metrics of technologies. Such tools are further discussed in Chapter 5 and have now gained wide application. By such measures Synbio has a key advantage in comparison with conventional engineering techniques such as nanotechnological robots in its source materials. While mechanically engineered robots require physical materials and energy sources, Synbio is built from existing biological materials and thus has an evolutionary 'edge' through molecular complexity. It is also plausible to combine Synbio with conventional engineering techniques, leading to a fusion of tools towards more sustainable outcomes. Such an eclectic approach that considers connections between various processes also links with point 5, which exhorts us to be systems oriented in our methods. A systems approach connotes a network with 'nodes' of functionality such as different organs of the human body. Any medical intervention to heal one organ must consider its impact on other parts of the body. Further, a systems approach allows for an early diagnosis of unintended consequences of technological interventions while also leveraging existing strengths of network nodes.

The final point in the technology assessment rubric considers the importance of governance through accurate information exchange. Various international mechanisms such as science panels have played such a role in the context of international environmental treaties. In the context of Synbio, the Convention on Biological Diversity (CBD) has at its service the Intergovernmental science-policy Platform on Biodiversity and Ecosystem Services (IPBES). There are also various ad hoc advisory panels which can be set up to consider particular technologies.

An important recent case in point with Synbio is 'gene drive' technologies that were evaluated by the CBD in terms of risk–reward trade-offs. Gene drive systems allow for the genetic modification of entire populations *in situ* (within the ecosystem) by releasing just a few modified individuals of that species. For example, a mosquito with a gene that would prevent the transmission of the parasite causing malaria could potentially be spread in an ecosystem to eradicate the disease. Activists had called for a 'moratorium' on gene drive technologies to assess the risks, but the scientific consensus favoured continued advancement of the technology and in 2018 the CBD member states rejected calls for a moratorium. Effective monitoring of the release of gene drive organisms was noted. The potential for these technologies playing a positive role in sustainable health outcomes through disease eradication outweighed the objections.

While constructive technological mechanisms are devised, there will still need to be social innovations to address key sustainability challenges. In the absence of risky technological interventions in the brain proposed by futurists, changing human behaviour requires a system of incentives and disincentives that we have come to calibrate through the contested domain of economic markets.

Economic development and ecosystem services

Economics is often defined as a study of how humans manage scarce resources, so it is highly relevant to questions of sustainability. However, conventional economics remains centred on paradigms of growth, treating environmental factors as 'externalities' to the core incentive calculus of markets. Still, there are a range of intervening factors which need to be considered even if they are not captured by the conventional market. Indeed, the goal of an economy which aspires for sustainability should be to find ways where possible of 'decoupling' localized growth from ecological harm.

The narrative around decoupling has dominated the work of the United Nations International Resource Panel. Ideally, we would want a society to work towards absolute decoupling, whereby energy and material usage as well as adverse environmental impacts would decrease as the economy grows. However, what is more plausible is relative decoupling, where resource usage and ecological impact increase with growth but at a slower pace and hence there is still potential for irreversible harm. Figure 11 shows the various forms of decoupling as they are considered in policy scenarios.

The figure shows both economic output variables and subject indicators of well-being such as 'happiness'. In the relative decoupling portion of the chart, there is continued resource depletion as economic output and well-being rise. In the second phase of the chart, resource usage per capita starts to fall while well-being increases—this can happen due to massive efficiency gains while reducing aggregate consumption per capita.

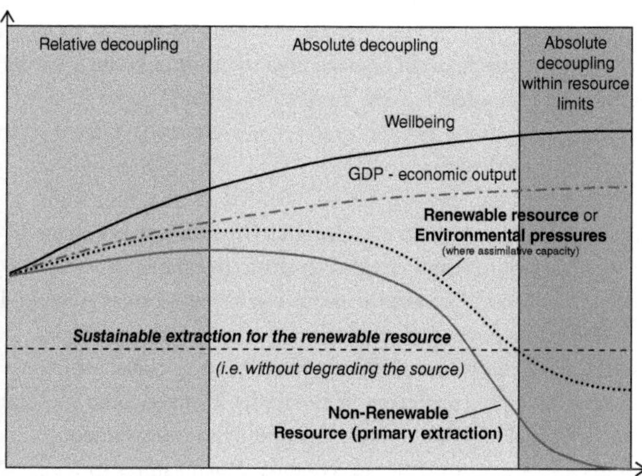

11. Various resource-decoupling frontiers from the Institute for European Environmental Policy.

In the final phase of the graph, the resource usage per capita goes below some measure of carrying capacity threshold. That is to say that the resources can be replenished—in the case of minerals this could be through recycling or a circular economy, and in the case of renewable resources this could be through natural regeneration.

The cyclicality of natural systems was also proposed in economic terms by the Russian scientist Nikolai Kondratieff, who suggested 'long waves' of economic growth driven by basic innovations and following a trajectory of expansion, stagnation, and recession. Starting from around 1780 until about the year 2000 there were five such observable phases or cycles with 40–60 years' duration which could be delineated as follows:

A) steam engine
B) railway and steel
C) chemistry and electrical engineering
D) petrochemical and automobiles
E) information technology.

Perhaps a sixth cycle could be suggested in terms of biotechnology, genetics, and nanotechnology. Unlike the delineation of 'four industrial revolutions' (steam, electricity, computing, and networked globalization) from the World Economic Forum, the Kondratieff cycles were much more empirically determined in terms of economic growth and decline. A next phase of technological optimism to deal with planetary challenges may occur in the form of geoengineering approaches such as solar radiation management. However, in a complex socio-ecological system predicting future Kondratieff cycles from considering past cycles can be highly problematic, no matter how good a fit the past may have provided. The natural cycles which sustain planetary systems also serve humanity in multiple ways. When we talk of 'primary' sectors in the economy, the term connotes natural capital, which in turn gets converted to other productive forms of capital in a modern economy. Nevertheless, the primacy of the natural capital in providing services for humanity is often occluded by the ways in which the extraction process may harm a natural system's long-term viability.

Natural systems provide essential life-support services for humanity that are often taken for granted. Clean air and water are fundamental ecosystem services that arise through natural processes. Recent efforts at documenting and even calculating the value of these natural capital services have become an important endeavour in sustainability science. These 'ecosystem services' can be quantified in financial terms and have led to the development of software packages that can undertake such calculations with an input of key natural capital indicators. The Integrated Valuation of Ecosystems Services and Trade-offs (InVEST) suite of models is provided open source by Stanford University's Natural Capital project and widely used. The United Nations has also developed a 'System of Environmental Economic Accounting' (SEEA) which utilizes the Artificial Intelligence for Environment and Sustainability (ARIES) modelling platform. Such tools have now gained mainstream traction in economic development analytics. The World Bank launched the 'Wealth Accounting for Valuation of Ecosystems Services' (WAVES) programme and used it in 2015 to consider biodiversity conservation programme benefits around a mining investment project in Madagascar.

A novel method to consider ecosystem services within a social context of sustainability was conducted in South Africa with the direct use of six provisioning services. These services were fresh water from a natural source, firewood for cooking, firewood for heating, natural building materials, animal production, and crop production. Based on a cluster analysis, the authors identified three distinct ecosystem service bundles that represent social–ecological systems characterized by low, medium, and high levels of direct ecosystem service use among households.

Two bundles were noted as 'green-loop' transition and 'red-loop' systems. Rural agricultural or 'green-loop' systems were defined by high direct dependence on local ecosystems, and little or no external economic interface through which to secure natural resources. In such systems there is direct feedback between human well-being and the degradation of the environment. On the other hand, in urban industrialized or 'red-loop' systems, almost all individuals in society secure their basic needs for food, water, and other materials through markets supplied by distant ecosystems, which leads to a social order that is detached from its local environment. These two system types face very different sustainability challenges.

In the green-loop system, the challenge—especially with demographic pressures—is to avoid a 'green trap' of ongoing poverty and excessive local degradation of ecosystems. In the red-loop system the challenge is to avoid over-consumption resulting from increased affluence and diminished human–environment connections, leading to over-exploitation of multiple, distant ecosystems, or the so-called 'red trap'. Identifying localities that are in these different social–ecological configurations, or in transition between them, is therefore essential in crafting policies that can calibrate resource usage with human well-being challenges in different areas.

Mapping such systems leads to integrated spatial units that differ from systems identified by additive combinations of separate social and biophysical data sets—thus providing an emergent 'socio-ecological system'. The distribution of localities is mainly determined by social factors, such as household income, gender of the household head, and land tenure, and only partly determined by the supply of natural resources.

There is abundant use of complex modelling software for optimization analysis and forecasting for various development trajectories. The Organization for Economic Cooperation and

Development (OECD) has harmonized the most widely accepted global economic and environmental change models of long-term development–environment linkages in their OECD Environmental Outlook series. The output reveals that economic growth for the foreseeable future will primarily occur in developing countries and have serious environmental implications. The contributions of environmental impact to development would rise in these poorer nations, particularly with reference to deterioration of air and water quality. While there would be 'conditional convergence' towards measures of economic productivity worldwide, this may not translate into improvements in quality of life or in the ecological condition.

For instance, the model captured data from 6,000 major rivers worldwide and the analysis showed that India, China, and Africa would account for almost half of all the water-induced soil degradation, and around one-third of all anthropogenic nitrogen loading into river-ways by 2030. Thus, forecasting models project that further economic growth in developing countries is likely to substantially worsen pollution levels in a 'business as usual' scenario, so pollution control mechanisms would be needed to mitigate these impacts.

The International Institute for Applied Systems Analysis (IIASA) has also developed Shared Socioeconomic Pathways (SSPs) scenarios for modelling energy and climate over a timescale of a century. These models are now widely used in sustainability science to examine the set of challenges humanity will face to adapt to global environmental change under different social, economic, and environmental conditions in the long term.

Quality of life and sustainability

The ultimate goal of human development practitioners is to improve quality of life and perceptions of well-being in populations. Conventional economic metrics of well-being and

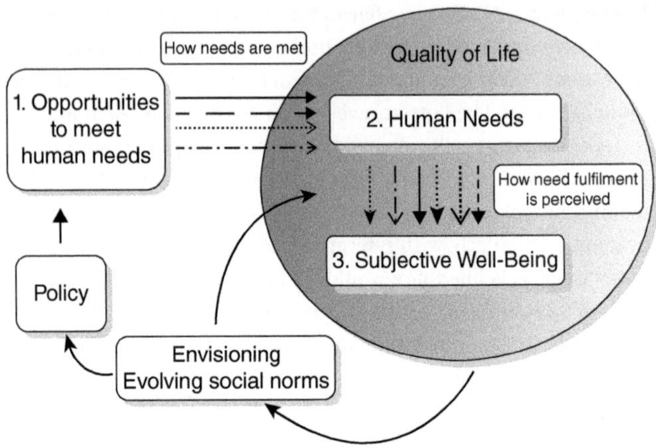

12. Relationship between quality of life and sustainability. The human needs aspect is resource constrained and hence linked to sustainability.

quality of life are increasingly being challenged with new indices that set forth, and challenge, the trade-offs between consumption and conservation. This section will discuss some of these indices and how they have gained traction worldwide. A simplified diagram that reveals these connections is provided in Figure 12 and shows the range of loosely (broken arrows) and tightly (solid arrows) coupled connections between human biological and social needs and various manifestations of 'subjective well-being' which are determined by individual human beings. Opportunities to meet these needs are also loosely and tightly coupled.

A key feature of subjective well-being connects to our perceptions of ownership and possession of a resource. Private property was deemed a solution to the 'tragedy of the commons' but has become an end in itself. Renting versus buying products could be a mechanism for mitigating resource intensity of usage. This can also shift responsibility for increasing longevity of a product through servicing with the producer while maintaining quality for the consumer. Photocopy machines were among the first such products to move from purchase to leasing arrangements when Xerox and other companies realized the value proposition of servicing as a means of reliable income while also reducing production costs. Sharing accommodation with family members, particularly by students, has been an important outcome of the pandemic. There were significant carbon reductions simply from young adults moving back home with their parents, and housing and office demand diminished. The long-term implications of 'Work from Home' will need to be evaluated at multiple levels in terms of productivity, quality of life, and ecological impacts. However, there is little doubt that a well-managed economy which has options for shared energy and material can contribute towards more sustainable outcomes.

Objective and subjective measures of well-being have also been operationalized in sustainability metrics such as the 'Genuine Progress Indicator' (GPI). This is a composite measure which has been proposed by ecological economists as an antidote to the dominant use of the Gross Domestic Product, which only measures quantity of economic output flows rather than their impact on quality of life and ecological variables. Since the advent of the GPI, which was initially piloted in the American state of Vermont, there are now several indices that measure well-being and have become part of the mainstream. The United Nations issues a 'World Happiness Report', the OECD has a 'Better Life Index', and several local jurisdictions have also started to use more nuanced measures of prosperity. Figure 13 shows the range of criteria which are measured in the GPI.

The key underlying question that remains challenging to answer definitively is the varied linkage between material consumption and subjective well-being. Although minimalist movements have gained traction through popularization of 'tiny houses' and a 'shared economy', there is still data to suggest that material

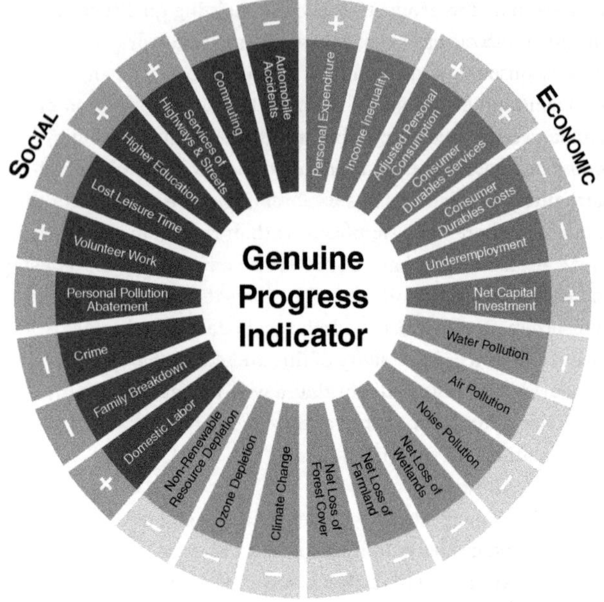

13. Metrics of GPI from the Gross National Happiness organization.

possessions and some degree of wealth alleviate both objective and subjective indicators of well-being.

In his acclaimed dialogue essay 'The Critic as Artist', Oscar Wilde stated that humanity likes to 'rage against materialism, as they call it, forgetting that there has been no material improvement that has not spiritualized the world'. This thread has also been picked up by many revisionist scholars of consumerism and material culture, who ask us to reflect upon the social ties created by gift-giving, the livelihoods created by the products which are produced as a result of conspicuous consumption.

In his provocatively titled book *Lead Us into Temptation*, James Twitchell tries to make the case primarily through observational analysis that consumer fashions and branding lead to bonding in an age of individualism. Renegade environmentalists such as Jesse Lemisch have also been irritated by many green activists such as Ralph Nader who they feel have 'turned their backs on people's reasonable and deeply human longings for abundance, joy, cornucopia, variety and mobility, substituting instead a puritanical asceticism that romanticized hardship, scarcity, localism and underdevelopment'.

British sociologist Daniel Miller believes that we can 'feel sympathetic to the dreadful plight of cosmopolitans who feel they have too many pairs of shoes . . . and they bought their child a present instead of spending quality time with them'. However, he considers it 'not acceptable that the study of consumption, and any potential moral stance to it, be reduced to an expression of such people's guilts and anxieties'.

Data on linkages between wealth, acquisitive impulses, and happiness are far more complex and deserve some careful consideration as we ponder material desires. 'Feeling good' has many technical connotations. Researchers tend to use terms such as 'quality of life', 'well-being', 'life satisfaction', and simply 'happiness' to define positive emotions that have physical and psychological aspects. According to social psychologist Sonja Lyubomirsky, the acclaimed author of *The How of Happiness*, there is a happiness 'set point' that accounts for about 50 per cent of individuals' happiness that is determined by temperamental factors (genetically determined). An additional 10 per cent is determined by social circumstances, and perhaps around 40 per cent is within our control in terms of behavioural choices. Consumption behaviour would fall within this 40 per cent of behavioural attributes that could potentially give us a higher sense of well-being.

Clayton Alderfer subsequently developed a non-linear version of Maslow's hierarchy of needs (noted in Chapter 1) in 1972, which was termed 'An Empirical Test of a New Theory of Human Need'. The key attributes of this model were *existence*, *relatedness*, and *growth* (ERG), which may be particularly useful in our attempts to understand both the positive and negative dimensions of our treasure impulse. Even though Alderfer had developed this model in the context of organizations, the applicability is much broader. The elegance of this simple model is that it shows how connections between material needs and more intangible aspects of well-being can be positive as well as negative, given the motivational direction.

Figure 14 shows an adapted framework of this model which synthesizes many of the themes covered in this chapter. There are some kinds of acquisitive impulses and material usage beyond basic needs which could have a positive impact on satisfaction,

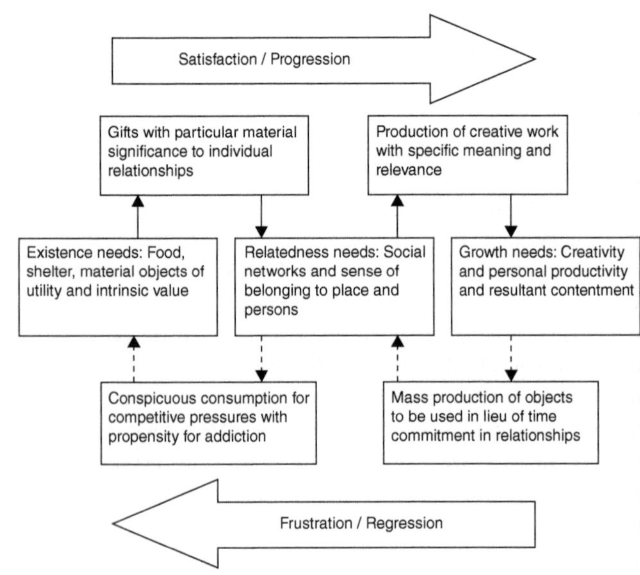

14. Existence Relatedness and Growth model of understanding human needs.

while there are many others that might not have that positive impact. Much depends on what the motivation and causal mechanism is for the consuming act and what meaning is derived from it.

In general, a certain amount of material wealth is important for individuals to be happy, but wealth is not a sufficient condition for happiness. Researchers are quite sensitive to differentiating 'income' from 'wealth'. The former category is an economic indicator of monetary flow in households and the economy at large, whereas the latter is an indicator of more lasting financial security. For our purposes of understanding material consumption, wealth is a better indicator, since it includes material assets and may account for factors such as debt and long-term financial security (and hence less worry).

However, for research purposes it is easier to get data on income and so most studies tend to focus on income, which can lead to what has been termed 'focusing illusions' by Nobel laureate Daniel Kahneman, who notes that humans exaggerate the contribution of income to happiness because they focus, in part, on conventional achievements when evaluating their life or the lives of others. We are thus better served in understanding the complexities of human interactions with material acquisition.

The mere act of seeking wealth can potentially increase well-being if we have a clear goal in mind but can be self-defeating if we do not have clear markers of achievement. One of the most important insights from psychological research on human well-being has been that the process of attainment is even more significant than the substantive goal itself. Most researchers agree that human beings tend to settle very quickly into a state of pleasure and then become 'bored' with their situation—a phenomenon called 'hedonic adaptation'. To remain satisfied, they need to be experiencing change, whether through the motivation of creative activity or through a positive quest. Remaining satisfied with life was thus likened by psychologists Brickman and Campbell to a 'hedonic treadmill'. The individual on the treadmill doesn't move forward but has to continuously strive to stay in a stable position—the connotation is thus rather pessimistic in their description of the phenomenon.

The negativity implied in this assertion has been widely refuted in the literature since their article was published in 1971. Surveys of happiness, despite their many methodological problems, have been found to show that individuals are happier in some circumstances and geographies than others. However, despite the various dissenting views, what has emerged is that positive pursuit is indeed an essential ingredient in well-being. Such pursuit must be anchored in a broader purpose and might not necessarily involve consumption. Building on the treadmill analogy, the individual may not be moving forward by some measures of subjective well-being but if the goal is framed in terms of physical fitness, they are very much on the right track.

In addition, it is essential to blend subjective well-being with objective criteria of human prosperity such as health, which usually figures as the most significant factor in well-being according to most surveys. Furthermore, given the transient state of human happiness at a given point in time, it is highly consequential to consider long-term impacts of consumption at other levels. The interface between consumption, the environment, and human well-being is now a subject of increasing academic inquiry.

Systems scientist Thomas Princen asks us to consider the trade-offs between efficiency of economies focused on large-scale production on the one hand, and the 'logic of sufficiency' on the other. If we consider longer timescales in our decision-making frames, it is more likely that we would act with what Princen terms 'ecological rationality'. We might begin to consider individual consumption decisions such as shopping choices with greater care. Just as nutrition labels made shoppers savvier with health-conscious shopping, perhaps we might develop an overall reduction on consumption for the sake of the planet?

Anxiety over whether or not humanity can achieve certain desired goals of sustainability has led to greater activism around 'sufficiency'. Such an approach suggests that ultimately the safest path forward is to hunker down and uncompromisingly consume less. Greta Thunberg's voyage across the Atlantic on a sail boat rather than by air exemplifies such an approach. Yet the counter-argument is that sufficiency can stifle innovation as well as human choice and pluralism in democratic societies. Navigating this debate between efficiency, innovation, and sufficiency constitutes the core set of conflicts facing sustainability professionals. Many of these debates require data to be resolved. Socially optimal decision-points can be calculated to some degree through use of techniques like 'multiple criteria decision analysis', with computer programs. Yet humans would still need to enter the weighting for various criteria in terms of their value for particular human populations.

The Chilean psychologist Manfred Max-Neef refined Maslow's hierarchy of needs to a matrix which did not assume value being attributed to one factor or another. He rejected the hierarchical principle as contrived, as needs could be equally important in parallel. Such an approach distinguishes basic human needs from economic goods. A conception of welfare based on this model challenges the assumed positive relationship between increased material consumption and increased satisfaction of needs, especially of non-material needs. The conventional economic approach assumes that subjective desires and preferences can be satisfied through consumer choices. However, this model questions the primacy of economic growth in the improvement of human well-being. In terms of sustainability, this allows for arguments that environmental factors should not be viewed as constraints on human welfare (Figure 15).

Fundamental Human Needs	Being (qualities)	Having (things)	Doing (actions)	Interacting (settings)
subsistence	physical and mental health	food, shelter work	feed, clothe, rest, work	living environment, social setting
protection	care, adaptability autonomy	social security, health systems, work	co-operate, plan, take care of, help	social environment, dwelling
affection	respect, sense of humour, generosity, sensuality	friendships, family, relationships with nature	share, take care of, make love, express emotions	privacy, intimate spaces of togetherness
understanding	critical capacity, curiosity, intuition	literature, teachers, policies educational	analyse, study, meditate investigate	schools, families universities, communities
participation	receptiveness, dedication, sense of humour	responsibilities, duties, work, rights	cooperate, dissent, express opinions	associations, parties, POW*, neighbourhoods
leisure	imagination, tranquillity spontaneity	games, parties, peace of mind	day-dream, remember, relax, have fun	landscapes, intimate spaces, places to be alone
creation	imagination, boldness, inventiveness, curiosity	abilities, skills, work, techniques	invent, build, design, work, compose, interpret	spaces for expression, workshops, audiences
identity	sense of belonging, self-esteem, consistency	language, religions, work, customs, values, norms	get to know oneself, grow, commit oneself	place one belongs to, everyday settings
freedom	autonomy, passion, self-esteem, open-mindedness	equal rights	dissent, choose, run risks, develop awareness	anywhere

*POW - Place of worship

15. Manfred Max-Neef's Model of Human Development.

Even with responsible material usage some ardent environmental critics of consumerism such as Juliet Schor would suggest that humanity is 'over-shopping'. If all the little additional trinkets and consumer goods that we buy are not definitively improving our quality of life, then why do we have to go on spending? Part of the answer may be that we have had greater access to credit, often to our detriment in terms of long-term planning, as exemplified by the financial crisis of 2008. With huge credit availability, we have been able to view opportunity for apparently immense wealth as far more reachable than in prior generations. Schor argues that material acquisition may provide the illusion of bridging financial inequality. However, the larger question remains to be addressed about what impact shopping has on the economy as a whole and on the livelihoods of those who produce the goods as a trade-off against environmental impact.

During economic recessions, there is often a call to increase consumer spending power. During the COVID pandemic, many countries made direct payments not only to ensure basic sustenance but also to keep businesses afloat and prevent a spiralling economic depression. The economic logic of this intervention suggests that spending would boost corporate incomes and potentially create more livelihoods. Shopping would most likely lead to economic growth and a moral argument could thus be made for it.

However, during wartime crises you can also have a call for reduced consumption to conserve essential goods for their most urgent needs to the military. During the Second World War famed public intellectual Lewis Mumford wrote of the need for reducing consumption as an act of patriotism rather than increasing consumer spending. The reason for this divergence of policy prescriptions in different crises is largely due to the underlying surplus or deficit of the natural resource base. We may have situations where there is a glut of produced crops or factory products due to errant planning decisions. Furthermore, it is also linked to what institutions we depend on to deliver natural resources and provide livelihoods. In the post-war economy there is far greater emphasis on entrepreneurship and private sector businesses to deliver economic development. Concomitantly, there is also far greater need for corporate practices to fall within the ambit of sustainability.

Corporations and sustainability

Corporations have often been described in organismic terms because of their capacity for impact on the environment and some of their behavioural attributes, such as resource consumption, waste generation, and growth. Corporate unions or mergers are frequently described in metaphors of matrimony. Occasionally, corporations may even exhibit a kind of amoebic procreation, such as the 'Baby Bells' telephone companies that emerged from the government-mediated dismemberment of the American Telegraph and Telephone Company (AT&T) in 1984.

In their landmark book *Organizational Ecology*, Michael Hannan and John Freeman first expounded on this framing of corporate behaviour. There was a Darwinian tone in their analysis that harks back to competitive markets being akin to natural selection processes. While many corporations may fit some aspects of this model, the growth of the mega-consumer-store, as exemplified by Walmart, challenges the model. When globalization and economies of scale allow for the consolidation of retail under one roof, the prospect for constructive competition can arguably be stifled. However, the concentration of consumer merchandise under a dominant retailer opens some remarkable opportunities for efficient environmental action.

Policy analysts have discovered that normative issues with future benefits such as environmental initiatives are usually most easily implemented through focusing on producers and service providers. Consumers are too capricious and often not adequately informed to even respond to basic market incentives in this regard. Furthermore, the speed of consumer response to technical information or prices might not be able to keep up with ecological stresses. High-income consumers who usually take on a very large share of product consumption might have highly inelastic demand patterns which would not respond to price fluctuations. How then might we spur positive change? Where in the food chain of material consumption might we get the biggest bang for the buck?

The usual response to spur producers to improve environmental performance has been regulation. Usually, such action is generated by litigation and/or serious confrontation. However, such action becomes increasingly onerous when the antagonist has market capitalization to rival several countries in the world. Consider the power that could be exercised by a company that accounts for 10 per cent of China's entire exports and where 82 per cent of American households have shopped at least once per year. Imagine the trickle-down to consumers and the trickle-back to suppliers if a company like Walmart were to consider themselves as agents of environmental action.

While Walmart has often been criticized for predatory pricing that decimates the fabled 'mom' and 'pop' shops, there is a surprising sustainability case to be made for a mega-store. Finding all consumer products under one roof reduces the need to travel to various destinations and also consolidates energy and resource usage more efficiently. However, the sourcing of the products is where diversification and allowance for local and varied production needs to be captured. While Walmart has kept prices low by buying in bulk along product lines, if we start to account for environmental costs the accounting may work out more favourably for local suppliers in some cases.

This is not to say that imports from China or other developing countries would disappear. Indeed, they should remain an important part of the consumer mix in developed countries to allow for wealth transfer to those most in need. However, accounting for environmental performance would assuage conditions in producing countries. Such producers are well suited to exercise influence given their huge buying power. Responsible actions by large corporations can thus have enormous leverage for transitioning society towards more sustainable practices.

Apart from influencing developing countries in their product development, multinationals can also have a tremendous influence on companies manufacturing particular kinds of products. Consider the lowly electric light bulb. The technology for more energy efficient bulbs (compact fluorescent lights or CFLs) has existed for decades and Phillips introduced the first such bulb to fit in conventional sockets in 1979. However, the manufacturing companies were not provided an incentive to produce more of these because the sellers thought they would be too expensive for the average consumer. Indeed, the incandescent light bulb was several-fold less expensive and until adequate regulator pressure and subsidies were provided a market for CFLs seemed illusory.

In December 2005, Walmart organized a meeting with General Electric Executives and invited various academic experts to join in the effort in making an ecological and economic case for CFLs. While initially the bulb manufacturers were reluctant, they eventually agreed to invest in this area because of Walmart's tremendous purchasing power. In addition, quality control and labelling criteria for bulbs helped consumers in making an informed decision about CFLs being a better economic and environmental choice.

Within two years of this partnership Walmart surpassed its goal to sell 100 million CFLs by the end of 2007. Over the lifetime of the CFLs, these energy-saving bulbs had the impact of taking 700,000 cars off the road, or conserving the energy needed to power 450,000 single-family homes. This scale of impact was arguably achieved with far greater efficiency because of the economies of scale that corporate sustainability efforts can provide.

There are legitimate concerns by activists that many corporate sustainability efforts can be a form of 'Greenwash', whereby labour practices or more systems-wide changes in procurement and sales can be eclipsed by well-marketed niche projects. However, such concerns can be addressed by having more robust audit and certification systems for sustainability. An industry of 'sustainability standards' has also developed in recent years to benchmark performance metrics in companies. Here too scrutiny is in order, to ensure the audits are done transparently and effectively.

A key concern with the dominance of private profit-driven corporations in shepherding sustainability is the timescales on which their incentive structures for performance are aligned. For publicly traded companies with stocks, there are short-term targets of creating shareholder value. Investors are generally interested in getting fast returns and less concerned with the long-term implications of those gains on natural resources being depleted from the planet because by then they may have already cashed in their profits. The next chapter considers how some of these natural processes and their triggers operate far from the orbit of market mechanisms and how this distance poses a challenge for sustainability governance.

Chapter 4
Tipping points and resilience

Why has Antarctica been at the heart of global environmental change research when it does not provide any direct resources to markets, nor is a viable habitation for human populations? Why would an uninhabited continent with potentially many natural resources have been set aside for science at the height of the Cold War through the Antarctic treaty system? These questions, which often arise in sustainability science lecture rooms, become even more salient as climate change leads to greater access and more liveable conditions on the continent.

The Antarctic treaty negotiations were motivated by the global environmental research community in the International Geophysical Year (1957–8). There was a stark realization that the sheer scale and unique features of the continent in its steady state were essential for maintaining the 'Goldilocks zone' of habitation for the rest of our planet. Inherent to this realization is the notion of 'global tipping points'. With any balanced system like a see-saw or a ball on a hill, there is a point at which the system 'tips over' into a different state. Natural systems exhibit such inflections in myriad ways. In a physical system it can involve a change in phase—water turning to ice once the temperature tips below 0 °C under normal pressure, for example. The phase change is spontaneous as the temperature tips. Furthermore, many tipping points in nature are connected, and this can lead to a domino effect which can be irreversible for all practical purposes.

Antarctica has several protective ice shelves that fan out into the ocean ahead of the continent's constantly flowing glaciers, slowing the land-based glaciers' flow to the sea. But those shelves can thin and break up as warmer water moves in under them. The breaking up of ice shelves can expose towering ice cliffs that may not be able to stand on their own. There are two potential instabilities at this point. Parts of the Antarctic ice sheet are grounded below sea level on bedrock that slopes inward towards the centre of the continent, so warming ocean water can eat around their lower edges, destabilizing them and causing them to retreat downslope rapidly. Above the water, surface melting and rain can open fractures in the ice. When the ice cliffs get too tall to support themselves, they can collapse catastrophically, accelerating the rate of ice flow to the ocean.

The Antarctic ice sheet is our planet's largest land ice reservoir (equivalent to 57.9 metres of Global Mean Sea Level) and its ice loss is accelerating. Extensive regions of the Antarctic ice sheet are grounded below sea level and susceptible to dynamical instabilities that can cross a tipping point of rapid retreat. For an ice sheet to not contribute to sea level, there must be a balance between the mass lost—through melting at the ocean and the breaking off, or 'calving', of icebergs—and the mass gained through snowfall. In a marine ice sheet, such as West Antarctica, the ice rests on ground below sea level. Here, retreat of the ice sheet can become self-sustaining through a 'positive feedback' system which can lead to marine ice sheet instability (MISI). This concept is particularly relevant in the context of Antarctica, where large portions of the ice sheet are grounded below sea level. If certain thresholds are crossed, it could trigger the rapid retreat of these ice sheets, contributing to accelerated sea-level rise.

The Antarctic is particularly vulnerable to MISI, but the positive feedback that makes this process a tipping point is driven by internal ice dynamics and is sensitive to both local and far-field conditions. This makes it difficult to classify whether a particular glacier on the continent has crossed a tipping point or not. Tipping points like the MISI are an example of a 'hysteresis loop', where the response of a system to a change in conditions depends on the history of that change. Such systems are non-linear and very difficult to model.

Upscaling the Antarctic tipping point to broader global environmental change phenomena is useful in various versions of what is referred to as a 'hothouse Earth' scenario. Figure 16 shows such a scenario across geological time and the natural variation range that has occurred within human history. One of the defining features of what is being termed the epoch of humanity (the Anthropocene) is our ability to alter the natural cycle.

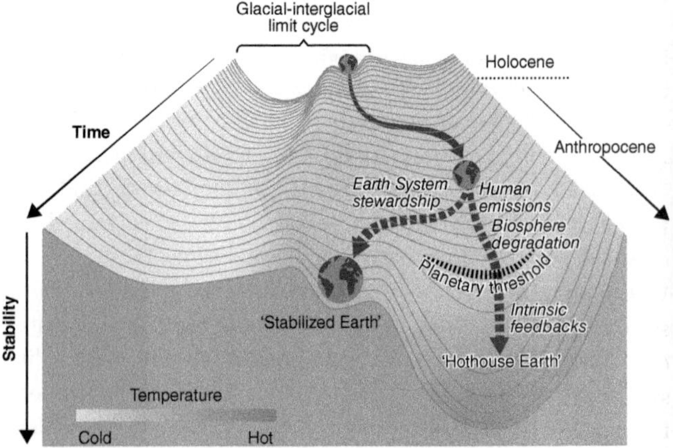

16. Global climate tipping point leading to a 'hothouse Earth' stability—a planetary threshold being crossed towards a stable but less hospitable world for humans.

Given the gradient of global environmental change, we cannot assume a return to an earlier state because the descent following the tipping point makes it far more difficult to reverse. In physics this phenomenon is also referred to as 'breaking of symmetry'. The threshold at which a 'bifurcation' may occur based on even a slight perturbation is the functional 'tipping point'. While many natural processes are reversible, there are also limits to resilience and recovery beyond critical points of inflection. Recent research has also shown that what we may perceive at macro-scales as reversible processes are simply a statistical manifestation of different irreversible processes occurring at various levels. For example, previously, a single enzyme was believed to catalyse both the forward and reverse chemical reactions, but more 'fine-tuned' research has found that two separate enzymes of similar structure are typically needed to perform what results in a pair of irreversible reactions in different directions.

However, the challenge of predicting and controlling such processes is that they are inherently 'chaotic' and hence the timing of such processes is highly difficult to predict. Furthermore, there are a series of such tipping points which could be interacting together.

The effect of cascading tipping points

Recent research analysing 30 types of natural phenomena encompassing physical climate and ecological systems—from West Antarctic ice sheet collapse to switching from rainforest to savanna—indicated that crossing tipping points in one system can increase the risk of crossing tipping points in other systems. Such links were found for 45 per cent of possible interactions. For example, Arctic sea-ice loss is amplifying regional warming and Arctic warming and Greenland melting are driving an influx of fresh water into the North Atlantic Ocean. Such a flow may have led to a recent 15 per cent slowdown of the Atlantic Meridional Overturning Circulation (AMOC). The AMOC involves the movement of warm surface waters from the tropics towards the North Atlantic, where the water cools, becomes denser, and sinks to deeper layers. The cold, deep water then flows southward and eventually returns to the tropics, completing the circuit that helps redistribute oceanic temperatures with seasons.

Rapid melting of the Greenland ice sheet and further slowdown of the AMOC could tip the West African monsoon, triggering drought in the Sahel. AMOC slowdown could also dry the Amazon, disrupt the East Asian monsoon, and cause heat to build up in the Southern Ocean, with potential to accelerate Antarctic ice loss. The analysis of past climates shows global tipping points, such as the entry into ice age cycles 2.6 million years ago and their switch in amplitude and frequency around 1 million years ago, which models are only just capable of simulating.

Although not directly applicable to the present interglacial period, this highlights that the Earth system has been unstable across multiple timescales in the past, under relatively weak forcing caused by changes in the Earth's orbit. Now we are strongly forcing the system, with atmospheric CO_2 concentration and global temperature increasing at rates an order of magnitude faster than during the most recent deglaciation. Carbon dioxide is already at levels last seen around 4 million years ago—long before the advent of modern humans—and rapidly heading towards levels last seen around 50 million years ago, in the Eocene, when temperatures were up to 14 °C higher than pre-industrial levels. It is challenging for climate models to simulate such past 'hothouse Earth' states.

Systems ecologist Tim Lenton led a team of tipping point researchers to map out the linkages between the nine global tipping points that are most consequential (Figure 17). Only the

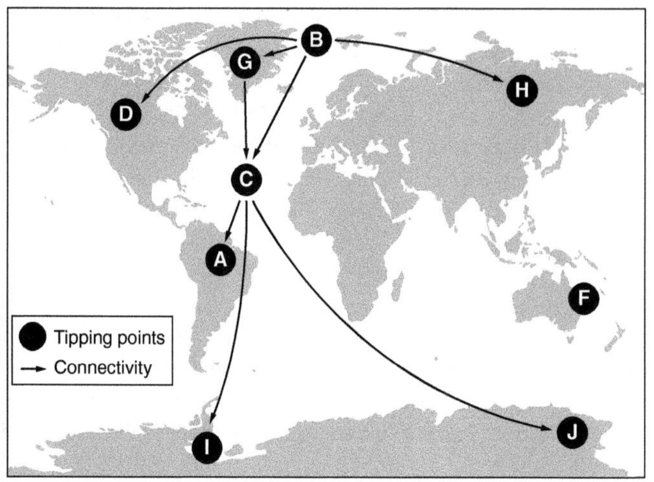

A. Amazon rainforest
Frequent droughts

B. Arctic sea ice
Reduction in area

C. Atlantic circulation
In slowdown since 1950s

D. Boreal forest
Fires and pests changing

F. Coral reefs
Large-scale die-offs

G. Greenland ice sheet
Ice loss accelerating

H. Permafrost
Thawing

I. West Antarctic ice sheet
Ice loss accelerating

J. Wilkes Basin,
East Antarctica
Ice loss accelerating

17. Linkages between global tipping points.

coral reefs tipping point does not have a direct connection with the other tipping points. Lenton and colleagues also considered how such cascades can reach the threshold of being labelled an emergency from a long-term sustainability perspective.

They define emergency (E) as the product of risk and urgency. Risk (R) is defined by insurers as probability (p) times damage (D). Urgency (U) is defined in emergency situations as reaction time to an alert (τ) divided by intervention time left to avoid a bad outcome (T). Thus:

$$E = R \times U = p \times D \times \tau / T.$$

The situation is an emergency if both risk and urgency are high. If reaction time is longer than the intervention time left ($\tau/T>1$), we have lost control.

Critics of such research metrics argue that the possibility of global tipping remains highly speculative and some tipping points which had initially been considered crucial, such as the Indian summer monsoon or methane hydrate release, are now less likely to occur. However, given the potential scale of impact and its irreversible nature, a planetary risk analysis of long-term sustainability of human populations under specific expectations must consider such tipping points. Much of this risk analysis needs to consider thermodynamics—the processes by which energy flows in systems and how that impacts the form and state of materials.

There are some thermodynamic processes which are functionally irreversible, such as gas suddenly expanding in an open system. Thermodynamic reversibility has statistical features in nature and needs to be controlled, so any discussion is best presented in terms of probability rather than certainty. When such processes occur naturally, we need to consider time as a variable as well. 'Time reversibility' is fulfilled if the process would happen the same way if time were to flow in reverse or the order of states in the process were reversed (the last state becoming the first and vice versa). The extinction of species and the ageing and death of organisms are examples of such probabilistically irreversible processes with implications for sustainability. Another example of irreversibility is materials such as plastics being deformed by energy such as heat.

If damaging tipping cascades can occur and a global tipping into another state cannot be ruled out, we should at least have a probabilistic analysis of impact versus likelihood, as shown in Figure 18.

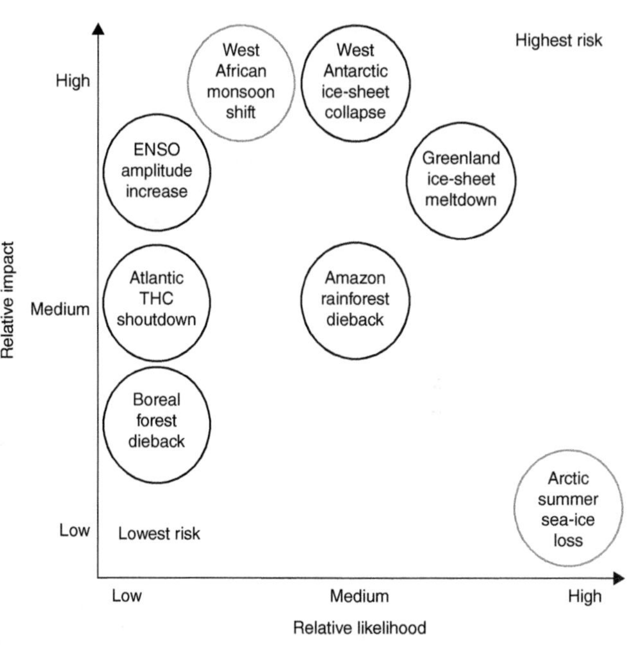

18. Relative likelihoods and impacts are assessed on a five-point scale: low, low-medium, medium, medium-high, and high. ENSO refers to El Niño Southern Oscillation, a cyclical process of warming of the Pacific; THC refers to the Thermo-Haline Circulation, a circulation pattern of mixing in the oceans due to differences in temperature and salinity.

Information on likelihood comes from review of the literature and advice from experts (faint rings indicate systems not considered in expert elicitation). The estimation of impacts is based on limited research and subjective judgement and is relative to the one system (bold ring), with multiple impact studies involved. Impacts are considered on the full 'ethical time horizon' of 1,000 years, assuming minimal discounting of impacts on future generations. (Note that most tipping point impacts would be high if placed on an absolute scale, compared with other climate eventualities.)

There are also examples of system reversals through simple shifts in net absorption and production of key compounds, which in turn can lead to a cascade of tipping points. Such shifts are already being empirically observed. For example, carbon dioxide production from burning of trees accelerates global warming additionally because the tree's ability to sequester carbon is lost and the resulting detritus can further release methane, a powerful greenhouse gas, from bacterial decay.

A team of researchers in Brazil looked at 600 vertical profiles of CO_2 and CO, carbon monoxide, which is produced by the fires, at four sites in the Amazon from 2010 to 2018. They found fires produced about 1.5 billion tonnes of CO_2 a year, with forest growth removing 0.5 billion tonnes. The 1 billion tonnes left in the atmosphere is equivalent to the annual emissions of Japan, the world's fifth-biggest polluter.

The positive feedback, where deforestation and climate change drive a release of carbon from the remaining forest that reinforces additional warming, and more carbon loss, was clearly observable from the data. The south-east Amazon has thus shifted from being a carbon sink to a carbon source.

A study conducted by the Potsdam Institute in late 2022 revealed that temporary overshoots of climate mitigation targets can increase tipping risks by up to 72 per cent compared with non-overshoot scenarios, even when the long-term equilibrium temperature stabilizes within the 1.5 degree range stipulated by the Paris Agreement. These results suggest that avoiding high-end climate risks is possible only for low-temperature overshoots and if long-term temperatures stabilize at or below today's levels of global warming. Such sobering findings lead us to consider if there are ways of controlling such processes at early stages of their movement towards a tipping point.

Controlling chaos in irreversible adaptive systems

One of the cardinal qualities of complex systems is that despite their seemingly unpredictable state of chaos, they tend to move towards 'attractors' and self-organize according to adaptive circumstances. Indeed, this property allows for sustainability to be manifest through multiple pathways towards some broad points of coherence. Complex systems theorist Stuart Kauffman has noted that sustainable societies live on 'the edge of chaos'. Both order and chaos occur in all functioning systems and they often oscillate between states for adaptation to occur in changing environments. In a physical science context, adaptation generally refers to the adjustment or modification of a system, organism, or process in response to changes in its environment. The term is used to describe how entities in the physical world evolve or change over time to better suit or survive in their surroundings. A surfeit of chaos leads to a loss of control potential. Too much order is equally unsustainable and leads to rigidity, a lack of variety, and decay. 'The edge of chaos' suggests states the system or object can move into—which could be a tipping point in either a qualitatively positive or negative direction.

Mathematical fractals are often used to describe scales in complex systems and these are also instructive for sustainability planning. There are structural factors which lead to similarities of scale such as 'power laws' in physical systems, which are manifest in observations such as the 'golden ratio' that is found in the shell of a nautilus but also in the spiralling arms of galaxies. Power laws imply a functional relationship between factors, where a relative change in one quantity results in a proportional relative change in the other quantity, independent of the initial size of those quantities: one quantity varies as a power of another. Working within the constraints of such power laws is likely to give us a greater chance of having a sustainable outcome. Here it is important to once again note the distinction between a steady state or equilibrium and sustainability, which can vary between scales.

Economics Nobel laureate Thomas Schelling developed these ideas in his classic book *Micromotives and Macrobehavior*, where he noted that economic equilibrium for whale hunting (whaling) during the 19th century did not emanate from a stable population which was in harmony with the larger oceanic ecosystem. Rather the whale population may reach economic equilibrium when the remaining stock of these mammals, prized for their meat and blubber, are so few that whalers can hardly catch enough to be profitable in their business. The few whalers who might invest in the high stakes hunting are just enough to offset the new births in the small population.

Another striking environmental example provided by Schelling pertained to the 20 to 30 million buffalo which had once roamed the western United States at the end of the Civil War. Only the tongues and hides of these animals were marketable, and they carried a high price, but the 'externality' of this market equilibrium was 20 billion pounds of rotting meat in the wild within six years. For every five pounds of buffalo meat left on soil, a merchant got a penny for the hide. Had there been longer-term thinking about the transport infrastructure needed to bring buffalo meat to consumer markets, with appropriate advertising the animals could have been worth more as live meat than just a source of hide within 15 years. Yet, the hunter's equilibrium was in that moment, and there was no mechanism to claim a property right to future live buffalo.

Equilibrium analysis thus oversimplifies outcomes by exaggerating points on a graph and neglecting processes of adjustment that are dependent on a broader range of parameters. Schelling was a master of metaphor and his writings were made vivid through elegant examples of the complex concepts he analysed through game theory:

The point to make here is that there is nothing particularly attractive about an equilibrium . . . Unless one is particularly interested in *how* dust settles, one can simplify analysis by concentrating on what happens after the dust has settled . . . the body of a hanged man is in equilibrium when it finally stops swinging, but nobody is going to insist that the man is all right.

Equilibria are not inherently 'right', and they may also be far more complicated in temporal and spatial dimensions than we might otherwise realize. This is where chaos theory can help to move towards a better steady state, and finding means of controlling pathways of chaotic expression becomes increasingly important. One of Schelling's key contributions to economic theory was the observation that equilibria may persist for a range of changing variables. He does this through a more nuanced understanding of what economists call 'utility functions'. A utility function is a mathematical representation that assigns a numerical value to different bundles of goods and services, reflecting an individual's preferences or satisfaction. Utility functions for consumers may encounter a rapid change only after certain thresholds are reached.

The salience of thresholds had its roots in the ground-breaking work by the systems scientist Herbert Simon in his questioning of the traditional view of economists that consumers maximize their utility during the 1950s. Simon recognized that rational choice had its limits and was 'bounded' by the information available to consumers and also their preferences for satiation. This insight added complexity and texture to our understanding of utility functions for consumers and ushered in the subfield of behavioural economics.

Around the time that Simon was developing his views on bounded rationality, computer science was reaching maturity as well and many of his ideas helped to nurture the field of Artificial Intelligence (AI), which is now considered an important means by which we might be better able to 'control' chaotic systems and move them towards more sustainable outcomes.

Simon's greatest contribution to social science was his willingness to embrace nuance and always question the contrived elegance of equilibria. In this spirit he coined a new term, 'satisficing'—a portmanteau of satisfy and suffice—to describe human behaviour in the context of bounded rationality. Unlike utility-maximizing options, satisficing options were those which are suboptimal and dependent on a process of alternative searches until a threshold is met for decision. For example, if you are searching for a restaurant, instead of meticulously researching every eatery to find the absolute best one, you are likely to choose a 'satisficing option' which you know is reasonably good and meets your basic criteria. Simon's realization that mathematical optimization was often limited in human decisions, because of 'computational intractability' and a paucity of information, transformed social science.

A year before Simon's death, in 2000, Carnegie Mellon University hosted a remarkable symposium titled 'Earthware' to consider how the metaphors of 'software' and 'hardware', as the architecture of computational order, could be applied to the Earth. The take-home message was that to control tipping points we need to operate with assumptions of 'bounded rationality' and also employ computational methods in developing 'smarter' human infrastructure.

Complex Systems theories help us understand how you can sustain a system over a long period of time, by achieving a workable balance of order and chaos, by working with attractors, by looking at multiple levels of a project, and by expecting the unexpected. In all systems, there are a whole lot of elements working randomly, but interconnected within scales and across scales. These connections constitute a network which in turn allows us to consider mechanisms for more effective control of the system.

Emergent properties imply 'non-equilibrium', which in turn can lead to irreversibility and the emergence of complexity. The next question to ask is, how does complexity impact resilience and, as a consequence, sustainability?

Complexity provides avenues for adaptation, as does mutation and natural selection. However, certain tipping points can mitigate adaptive responses by creating energy gradients which are difficult to surmount. Take the example of how certain organisms like cockroaches are able to adapt far more quickly to a range of ecological disruptions, while megafauna such as tigers or elephants cannot do so. Similarly, some social systems and cultural norms are more resilient and adaptive than others. Adaptive ability in social systems is often determined by diversification of the natural capital on which the social system is dependent.

After a major shock or disruption there is often an immediate impulse for humans to 'bounce back'. However, a dynamic view of sustainability that considers multiple trajectories emanating from complex systems analysis suggests that we should instead consider ways of 'bouncing forward'. The COVID pandemic can be used as a parable for considering how human systems, from prevention of disease to urban planning, can learn and redesign systems which are more resilient. The diagnostic power of disruptions in laying bare vulnerabilities in systems was evident from the research on COVID as a zoonotic disease (jumping from one species to another). The virus had survived in its host reservoir of bats or other mammals in the wild for generations but at some point there was a spill-over to human populations. The initial efforts at containment failed and once we crossed a tipping point of contagion the disease became a pandemic with natural immunity generation or vaccination being the only solution. However, just as tipping points explained the contagion, they could also explain a solution that would lead us to 'herd immunity' or a stable seasonal cycle of the disease.

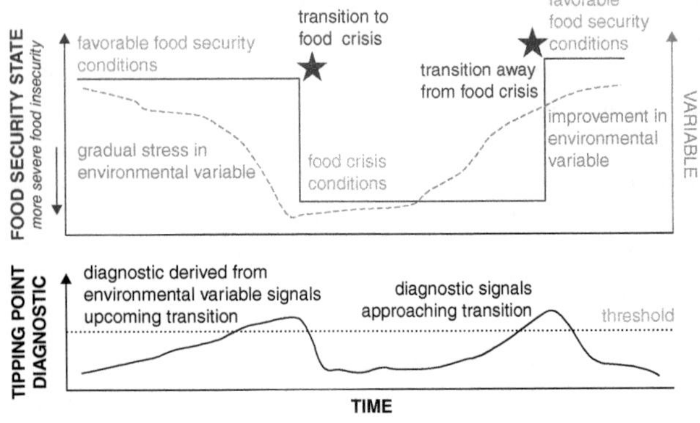

19. Applying tipping point theory to improve early warning drought signals for food security.

Tipping points can thus be just as much explanatory of solution success towards more sustainable goals as they are of diagnostic decline towards a crisis. Figure 19 shows how tipping point theory could be applied in the context of a drought crisis management system.

The authors of this study used seven variables derived from satellite observations relevant to drought and famine: precipitation; groundwater; snow; soil moisture; evapotranspiration; vegetation health; and chlorophyll fluorescence. Tipping point theory is thus a plausible heuristic to develop diagnostics that can warn of an impending crisis *before*, as opposed to during, the event. However, practical application of tipping point theory to sustainability planning requires large and high-definition data streams (in time and space). These in turn require major computational power which is at the heart of planning 'smart' and more sustainable design features for human habitation.

For lasting change in systems of human habitation, it is essential to engage experts from the social sciences and humanities.

The good news is that the idea of tipping points also has traction in a social context and is now being applied to consider policy interventions that can lead to lasting social change. Such 'social tipping points' occur when a small quantitative change can trigger rapid, non-linear changes. As with physical tipping points, such changes occur by self-reinforcing positive-feedback loops. Behavioural change can then become unavoidable for the larger population even if it may disagree with the underlying tenets. Such a process can thus lead to an irreversible and qualitatively different state of the social system.

Consider the switch from plastic bags to reusable bags in retail. There was immense resistance to such a change in many communities, but through regulatory changes, as well as market savings considerations from wholesale retailers such as Costco, plastic bags have quickly lost appeal in many communities. Yet the data on this change is complex, as the tipping point could not be reached if there was simply a nominal charge made on the bag, as was initially tried by retailers like Target. Might the same be possible with uptake of electric cars, whereby a confluence of regulations that necessitate phase-out over time of petrol vehicles, as well as availability of charging stations, may lead to an accelerated voluntary tipping point for consumers? This is what countries such as the UK are hoping for with their current EV policies. The key difference here with the plastic bags example is that the uptake of electric cars is also materially dependent on the availability of batteries that need critical metals. Figure 20 shows the ways in which an authoritative study on social tipping points conceptualized their interaction with physical tipping points for decarbonization. The efficacy of such social 'weights' to lower the barrier for tipping over to a more sustainable state is dependent on the complexity of the system (depicted by multiple troughs and peaks).

The researchers also identify social tipping elements (STEs), which have a key characteristic: 'A small change or intervention in the

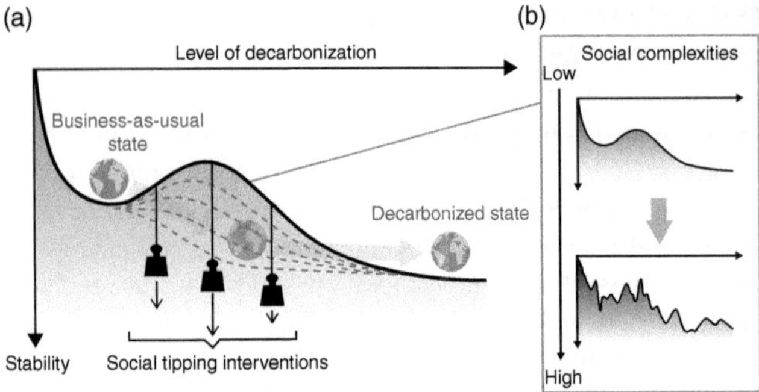

20. Social and physical tipping point interactions. If social complexities are high we get a more textured graph as shown in the side-boxes on the right.

subsystem can lead to large changes at the macroscopic level and drive the system into a new basin of attraction, making the transition difficult to reverse.' They also note the importance of social tipping interventions (STIs), which could be most effective in reducing barriers to change most efficiently. These STIs are strategic actions or policies designed to induce and accelerate positive and transformative changes in social systems. Such social tipping interventions aim to identify leverage points within a system where targeted efforts can bring about significant and positive changes. These interventions often involve a combination of policy measures, public awareness campaigns, and community engagement to create a transformative impact on a societal level.

Another means of considering social tipping points in the context of transport may be the uptake of bicycles and public transport. Copenhagen saw a rapid rise in bike use with infrastructure development and a 'herd effect', with young people wanting to have more physically active and sustainable lifestyles. Such scale effects of social tipping points are particularly significant in cities—which is where almost 70 per cent of the world's population is expected to reside by 2050. Developing infrastructure in cities that integrates social and physical systems will be key to reaching sustainability targets.

'Smart' infrastructure and sustainability

Cities are a microcosm for developing a sustainable society. Most of humanity is transitioning towards urban life as it is more efficient in terms of natural resource usage. Although the COVID pandemic and telecommuting technologies may have made 'Work From Home' more popular, cities continue to grow. While more people may end up living in picturesque locations, there will inevitably be clustering that occurs there as well in the same way as we have 'attractors' in complex systems. Tipping points would operate in such contexts too. Indeed, some of the earliest research on tipping points was done in the context of how population demographics can rapidly change in cities. In the 1950s, the term was used to describe how 'white flight' would occur in neighbourhoods when the number of Black families reached a certain percentage.

However, tipping point theory and the application of complex systems network analysis can now be operationalized to build more inclusive and sustainable cities, if we can appropriately channel data to make more effective decisions. The notion of 'smart' cities is linked to the feedback mechanisms that are similar to human intelligence. As the name implies, such habitations require artificial intelligence systems to be coupled with infrastructure. All such systems require massive amounts of data and a concomitant set of sensors and monitoring mechanisms, which can also trigger privacy concerns.

As a planned economy, China has excelled at grand infrastructure projects, and it has brought that approach to bear on ecological goals as well. The world's largest smart 'eco-city' is being developed by the Chinese government close to Tianjin, a 30-minute high-speed train ride south of Beijing. The site for the eco-city was chosen based on two key factors: that its development should occur on non-arable land; and that it would be developed near an area facing water shortage. The goal of the eco-city project is to find ways of using resource-scarce environments more productively, rather than optimizing particular ecological criteria by choice of location.

There are 26 key performance indicators (KPIs) which have been chosen to define and measure many of the social and environmental goals of the eco-city. These include 100 per cent potable water availability and zero use of bottled water; greater than 70 per cent native vegetation; and greater than 60 per cent recycling rate. Eco-cities need to be measured through such indices, which balance employment opportunities and housing availability in a particular area. Achieving equilibrium between job opportunities and housing availability is often seen as essential for sustainable and livable communities. The Tianjin local government has also committed to building an 'industrial ecosystem' that would create greater resource efficiency and exchange between various manufacturers and also green the supply chain at a global level.

Finding ways of reducing cost while staying within some parameters of green design has been challenging for architects working on the project. To attract a broad portfolio of professions some compromises have been made in terms of materials being used for construction. Nevertheless, the overall goal of scaling up ecological design and mitigating impacts on arable productive lands and finding non-conventional sources of water (recycled or through renewable energy-powered desalination) is admirable.

Continuing pollution in neighbouring areas may also hamper the efficacy of the eco-city. Yet, the lessons of Tianjin eco-city can certainly be used for a plethora of new urban development projects which will be inevitable alongside current demographic trends—particularly in Asia. China has also set up a broad range of sensors and data-gathering mechanisms which can provide feedback on improving efficiency of operations in the city. The project is being developed in partnership with the city-state of Singapore, which is known for its resource efficiency and innovation in managing a scarce resource base for development.

Singapore has also managed an ethnically diverse population through fairly stringent regulations on cohabitation and mixed income housing to mitigate gentrification and ghettoization. The country introduced an 'Ethnic Integration Policy' (EIP) in 1989 to ensure that public housing (in which over 80 per cent of the country's 5.7 million population reside) is inclusive and diverse. The EIP is implemented for all ethnic groups and places limits on the total percentage of a block or neighbourhood that may be occupied by a certain ethnicity. As soon as these limits are reached, no further sale of apartments to the affected group is allowed, unless the seller and buyer belong to the same ethnic group. Singapore has thus tried to engage in not just 'smart' technological development of its urban landscape but also a form of social engineering which has so far been able to keep the peace between its diverse set of ethnicities.

While the social outcomes of diversity and inclusion have generally been applauded, the technical side of such developments has led planners to raise serious concerns. The rush towards using big data to create a technologically sustainable tipping point has also come under considerable criticism—even from some engineers. Shoshanna Saxe, a Canadian systems engineer, notes that such smart cities would still have unpredictable vulnerabilities. She has argued that 'Tech products' age fast and are too reliant on sensors and reliability of data quality and processing speed. Furthermore, humans and indeed material flows are still needed to keep the cities running. If smart data identifies a road that needs paving, for example, it still would require construction workers to show up to build the infrastructure. Saxe succinctly states that rather than being driven by the allure of smart cities, which could tilt tipping points either way, we should instead be focusing on 'excellent dumb cities' that could be more resilient.

Instead of such smart cities predicated on synthetic technological systems and AI, there are growing calls for 'biomimicry' and 'nature-based solutions' for more sustainable and resilient cities. Nature-based Solutions (NbS) are defined by the International Union for the Conservation of Nature (IUCN) as 'actions to protect, sustainably manage, and restore natural or modified ecosystems that address societal challenges effectively and adaptively, simultaneously providing human well-being and biodiversity benefits'.

The argument for such solutions in comparison with 'smart' solutions with sensors, data, and machine learning is that biotic systems can exhibit properties of 'antifragility'—disruptions and strains can often make them stronger. For example, muscles in animals gain strength when subjected to stressful exercise up to a certain point. Using the analogy of tipping points, we can say that biotic solutions can expand the surface at the crest of a tipping point so as to more effectively prevent a negative slide down the precipice.

Traditional architectural designs rely on natural solutions such as the use of specific botanicals for insulation or designing air circulation systems through architectural techniques which are more aligned with natural systems. A remarkable example of this approach is the Eastgate Centre in Harare, the capital of Zimbabwe. The seven-storey multiple-use building has no heating or air conditioning. It is among the largest structures in the world to only use natural air flow and traditional biomimetic design inspired by termite mounds for thermal control. Such an approach to sustainable design makes it much more resilient as there is no reliance on a particular energy source for delivering a preferred outcome.

Despite occupying just 2–3 per cent of Earth's land surface, cities are home to most of humanity. Today some 4.46 billion people, or 57 per cent of the global population, live in cities. By 2050, this figure is projected to reach 6.7 billion, accounting for over two-thirds of world demographics. Cities are thus the prototypes for planetary sustainability conversations. Their scale also makes them a highly efficient means of testing other features of sustainability such as the circularity of material flows, which are the focus of the next chapter.

Chapter 5
Renewability, circularity, and industry

Biotic systems appear to show circular patterns of resource flows which subsequently lead to the biogeochemical cycles we considered earlier. This has been a hallmark of sustainability discussions, where an attempt to mimic biological systems is celebrated. Sustainability discourse is often peppered with the need for 'circularity'. For natural resources, we often use the term 'renewable' to indicate a preferred attribute of agricultural or forested systems, as opposed to mining, and the term 'renewable resources' appears normatively, but from a sustainability science perspective the underlying energy and material flows need to be considered before we become complacent.

Minerals are the ultimate primary resource, as they provide the chemical elements needed not only for the built environment but also for biotic systems. And chemical elements are fundamentally renewable, because they can cycle back interminably through a system, unless there is a nuclear reaction that converts one element into another. The question, then, is not whether minerals extraction is renewable but rather what amount of energy is needed to bring the mineral (and usually its derivative metal) back into functional use after a product has been consumed.

The work of Chilean systems biologist Humberto Maturana is illuminating here. In studying circularity of living systems, Maturana and his student Francisco Varela coined the term 'autopoiesis' to describe a network of processes of production (transformation and destruction) of components in living systems. The term is derived from the Greek words 'auto', meaning self, and 'poiesis', meaning creation or production. Autopoiesis refers to the self-creation or self-maintenance of living systems. Essentially, autopoiesis describes the ability of living organisms to produce and regenerate themselves continuously and autonomously. This concept emphasizes the dynamic processes of self-renewal and self-maintenance that living systems undergo to preserve their identity and structural organization. According to Maturana and Varela, an autopoietic system is characterized by a network of components (such as cells in the case of biological organisms) that interact in a way that enables the system to produce and maintain itself.

However, autopoietic systems can only function if they have a continuous supply of available energy and material. They create order through this process but are dissipative and also produce waste material as well, which in turn could be used by another autopoietic system as material or energy for their own subsistence. The interactions at different scales between autopoietic entities creates an ecosystem.

In 1986, NASA published an important report on planetary sustainability to consider how the impact of human activities would intersect with planetary processes (as noted in the Very Short Introduction on *Earth Systems Science*). A core component of this study was a diagram developed by atmospheric systems scientist Francis Bretherton which laid out the circularity which is possible between natural and human systems interacting together to support life on Earth. The diagram, which bears his name, is a key anchor in Earth systems science. The key features of this diagram are resonant with some of the themes presented in James Lovelock's Gaia hypothesis, which was often misinterpreted as suggesting that the Earth is a superorganism. Rather, what Lovelock and other scientists interested in planetary systems were suggesting was that Earth systems can have specific feedback loops that perpetuate a system that is supportive of life.

Robert Kates, William Clark, and 12 other scholars attempted to harmonize some of the key features of sustainability from a scientific lens in a short article in the journal *Science* in 2001, in which they posed the following seven questions that needed to be considered by industry and governments alike:

- How can the dynamic interactions between nature and society—including lags and inertia—be better incorporated into emerging models and conceptualizations that integrate the Earth system, human development, and sustainability?
- How are long-term trends in environment and development, including consumption and population, reshaping nature–society interactions in ways relevant to sustainability?
- What determines the vulnerability or resilience of the nature–society system in particular kinds of places and for particular types of ecosystems and human livelihoods?
- Can scientifically meaningful 'limits' or 'boundaries' be defined that would provide effective warning of conditions beyond which the nature–society systems incur a significantly increased risk of serious degradation?
- What systems of incentive structures—including markets, rules, norms, and scientific information—can most effectively improve social capacity to guide interactions between nature and society towards more sustainable trajectories?
- How can today's operational systems for monitoring and reporting on environmental and social conditions be integrated or extended to provide more useful guidance for efforts to navigate a transition towards sustainability?
- How can today's relatively independent activities of research planning, monitoring, assessment, and decision support be better integrated into systems for adaptive management and societal learning?

These questions are all related to our quest to consider how best to transition from a linear economy to a circular one in which human systems emulate the cyclicality of natural systems. Among the great challenges of sustainability has been the way humans have been able to harness minerals and develop their own synthetic materials, particularly since 1950, during the Great Acceleration. Figure 21 shows the remarkable rise of concrete and cement as key materials that have allowed for an unprecedented expansion of human infrastructure and the rise of cities.

The rapid rise of the use of such materials has been among the key arguments made for labelling the current epoch since the Great

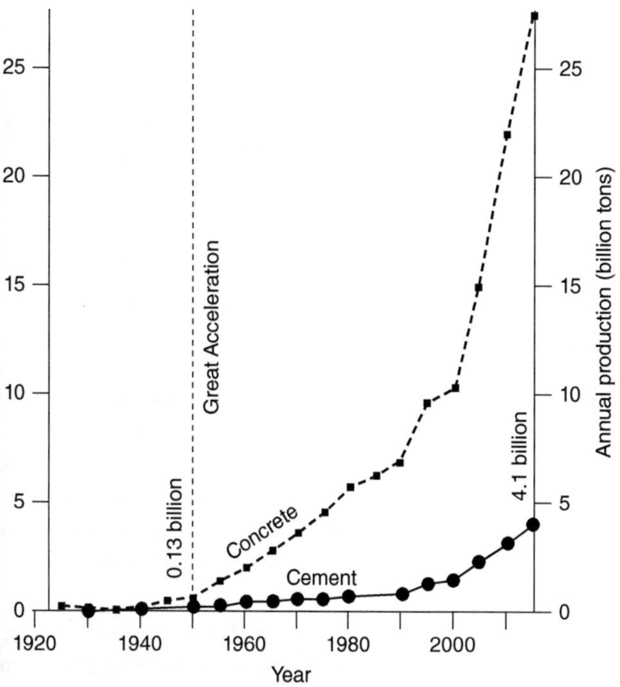

21. The rapid rise of anthropogenic minerals.

Acceleration as 'the Anthropocene'. Human industrial development has created ecosystems which have developed adaptive features that make them self-sustaining. Shipwrecks can become havens for marine life; fungi have evolved to feed on synthetic oils from leaking fuel tanks in the Antarctic; and birds have evolved to use urban skyscrapers as preferred nesting sites. These are all examples of how adaptation in complex systems can emerge. An important feature of this human dominance of life on Earth has also necessitated that we consider the importance of industrial systems as part and parcel of the larger 'natural' system. This has led to the development of a whole new field of academic inquiry, called 'industrial ecology', that is closely linked to sustainability science.

Industrial ecology and sustainability assessments

The field of industrial ecology developed out of a realization that industrial development was now a permanent part of the planetary system and we needed to consider synergies between natural and anthropogenic processes. The paradigm assumes that anthropogenic systems of production can mimic or harmonize with natural systems in terms of energy and material flows. On land, such interactions have more opportunities than they do in the oceans, since there are numerous nodes of human industrial activity. So it is easier to conceive of phenomena such as industrial symbiosis or eco-industrial parks on land than it is in the oceans where industrial activity is often more targeted, distant, and disconnected from other human activities. For example, some of the key human industrial activities in the oceans, such as long-distance fishing, shipping, or oil and gas extraction, have limited linkages to most people's homes or communities.

Let's consider the example of a mining project to explore how industrial ecology methods can be applied. In this context we could approach our analysis in two ways: local environmental resource usage in the production of the mine's output itself; and the ways in which the mineral products from the mine are used in society, which requires an evaluation of the entire downstream supply chain for the mineral. Let's look here at facets of both local and systemic industrial ecology in the context of sourcing metals from the oceans. Indeed, the local impacts of the extraction process inevitably get incorporated into systems-wide calculations. There is no clear, commonly accepted definition of industrial ecology due to different perspectives and techniques. Generally, however, industrial ecology is described as a field that combines economics and ecology to evaluate micro to macro behaviours in industrial systems based on observations of the behaviour of those systems. Some of the key features of an industrial ecological system are design for environment, dematerialization, life cycle analysis, and industrial symbiosis.

Design for Environment (*DFE*). As a starting point, the design of industrial systems to consider ecological factors requires us to go back to a basic thermodynamic evaluation. The system should have fundamental attributes of mitigating energy and material usage in delivery, mitigating waste generation, and subsequently facilitating the reuse and recycling of material inputs towards a circular economy. Design for 'disposability' has been encouraged for sanitation and hygiene reasons as well as for 'freeing up time' for more productive uses than cleaning and servicing used goods. By one estimate, articulated in the anthology *Natural Capitalism*, only 1 per cent of the North American material flow is still being used within products six months after sale and the rest is disposed. DFE attempts to reverse such trends by developing more resource-efficient products.

Dematerialization. Natural systems tend to develop efficient means of minimizing energy in their form and eventual systems-wide purpose. So industrial ecology practitioners consider ways of 'dematerializing' products (using fewer materials to serve the same purpose of the product) to create functional but resource-efficient goods. There is not much evidence that industries around the world are under dematerialization at present, despite individual products like phones and computer processors becoming smaller and more material efficient. The international extraction of six minerals (bauxite, the platinum group, magnesium, cobalt, molybdenum, and nickel) and the production of cement grew faster than GDP from 1960 to 2019. Industrial ecologists therefore aspire to systems-wide dematerialization, but are always having to also consider product quality, in engineering the product towards an optimal material need. The University of Cambridge 'Use Less' group led by Julian Allwood has pioneered various strategies on dematerialization. For example, additive manufacturing technologies, like 3D printing, allow for the creation of products layer by layer, often using fewer raw materials than traditional manufacturing methods.

Life cycle analysis (*LCA*). Much of industrial ecology discourse has developed around a range of methodologies that compare the system-wide impact of a product from its fundamental ingredients to its ultimate disposal and re-entry into productive use. Such 'life cycle analysis' can range from calculating ecological footprints to metrics on social cohesion. Resources which are conventionally considered 'renewable', such as water, food crops, or trees, also have an embedded energy footprint in terms of their availability for human use. Land itself is largely fixed and its functionality depends on nutrient flows and replenishment of key minerals needed for the growth of agriculture or forestry. LCA aims to consider all these latent factors in comparing composite impacts across products to improve material choice in manufacturing.

Industrial symbiosis. Just as organisms in the natural world can benefit from each other through symbiotic partnerships, industrial entities can also find pathways to more efficient metabolism across their 'ecosystem'. The term 'industrial symbiosis' describes a network of diverse organizations which

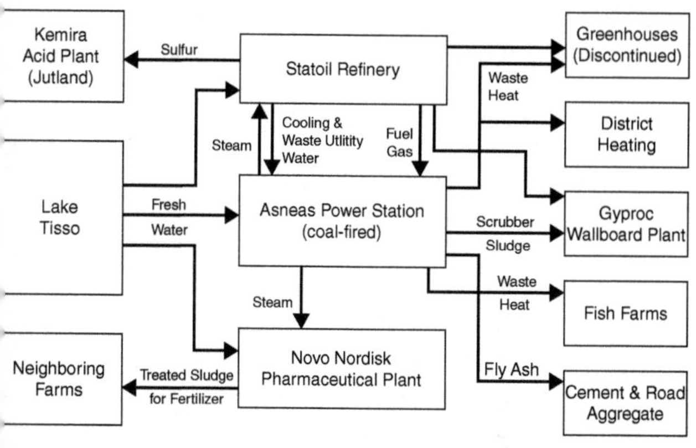

2. An industrial symbiosis system from Kalundborg, Denmark.

an foster innovative exchanges of material flows by using waste laterial and energy to generate productive usage. Eco-industrial arks, where one factory's waste is utilized by another, is an xample of such a process. Figure 22 shows one of the earliest ich parks in Kalundborg, Denmark, which has become mblematic of industrial ecology.

ndustrial ecology approaches play a crucial role by extending eyond the mere engineering aspects of a product. They involve onducting 'sustainability assessments' at various levels of the ipply chain. Our dependence on 'raw materials' establishes a irect link with the social dimension of sustainability. Mining, a particular, faces significant public disapproval, surpassing ven the tobacco industry. This disdain often stems from the ict that the environmental consequences of mining are not ypically experienced in the same location where the material's enefits are realized. Let us consider an example of how to onduct a 'sustainability assessment' at a mining project site, aking into account social impacts and benefits. It also ddresses the challenge of overcoming the 'not in my backyard' NIMBY) syndrome.

Example of a 'sustainability assessment'

To implement sustainability in major industrial development projects, there has been a growing interest in not just conducting the usual 'environmental and social impact assessments' but specifically 'sustainability assessments'. Extractive industries present a particularly challenging example of such an assessment. How could a mining project conduct a 'sustainability assessment', when the extracted resource itself is functionally finite? Clearly a systems view of sustainability is required at multiple levels to conduct such an assessment rather than just focusing on the mineral reserve itself. Among the first and most successful sustainability assessments conducted for a mining project, which motivated an entire book on the topic, occurred in the 1990s in a remote part of far eastern Labrador, Canada.

The Voisey's Bay project was a major nickel mine that was to be developed by INCO corporation at a time when indigenous rights in Canada were gaining political salience. Two indigenous groups, the Innu and the Inuit, were involved in the project and the company committed to developing a 'Sustainability Assessment' and an 'Impact Benefit Agreement' with the community. An expert panel was convened to prepare the groundwork for the sustainability assessment and they laid out the following three parameters:

- preservation of ecosystem integrity, including the capability of natural systems to maintain their structure and functions and to support biological diversity;
- respect for the rights of future generations to the sustainable use of renewable resources; and
- the attainment of durable and equitable social and economic benefit for the communities involved.

Ensuring that the mine provided such a trajectory required a longer vision of economic and social 'multiplier' effects. Turning the natural materials we get from mining (a form of 'natural capital') into other lasting and renewable investments is important. These investments can continue to give us benefits even after the mine is no longer active. It's not just about the technical details; people in the community must agree with the process for it to be socially sustainable. When the community, the company, and the government all agreed on how the project should be done, they put it to a vote for the whole community. It's impressive that almost 70 per cent of the community participated in the vote, and about the same percentage voted in favour. This strong support from the community added even more credibility to the project.

The Voisey's Bay case study is a good example of how 'weak' and 'strong' sustainability arguments can be operationalized. Environmental economist Sir Partha Dasgupta has noted that in both weak and strong cases, 'capital', whether natural or human, is carried on throughout the generations. It is the 'type' of capital that is important. Strong sustainability implies carrying on to the next generation the same amount of natural capital, with increases in human capital over time; whereas weak sustainability implies a declining natural capital over time while human capital increases. Table 1 provides a typology of how this kind of approach should be considered in broader debates about sustainability. Implicit in a technocratic approach is a worldview that people may not change their behaviour to meet ecological goals and hence we need to either engineer abundance or be willing to accept suboptimal outcomes. An eccentric approach is inherently more normative and aspirational.

lthough 'weak' may connote an inferior form of ustainability, such a presumption is only applicable if one onsiders natural capital to be inherently more valuable than other forms of capital. However, if we inject technological adaptation and consider a qualitatively different global

Table 1. Appraising weak and strong sustainability

	Technocentric		Ecocentric	
	Cornucopian	Accommodating	Communalist	Deep ecology
Sustainability labels	Very weak	Weak	Strong	Very strong
Environmental approach	Market-driven resource signals	Resource conservationist	Resource preservationist	Pristine restorationist
Type of economy	Neo-classical growth-driven	Green growth/circular economy	Steady-state	Degrowth

equilibrium for human well-being, then a weak sustainability outcome may be preferable because higher levels of per capita consumption could be obtained. The strong sustainability paradigm focuses on a replenishment of natural capital and maintaining its asset base at a certain level. However, such an equilibrium is more dependent on population stabilization as well as per unit consumption reduction.

Strong sustainability is thus more normative and requires far more individual discipline on the part of the consumer and parameters around what forms of innovation and lifestyle adjustments can be accommodated within planetary ecological parameters. Aspiring towards strong sustainability, while realistically making weak sustainability efficient in the extension of natural capital, is the core challenge for 'sustainable development'. The most common example of strong sustainability is an agroforestry system which is being organically managed for fruit harvesting. The natural capital is being regenerated in a biotic cycle. In contrast, a mineral extraction project which converts the mined metal into manufactured capital is an example of weak sustainability. However, the metal could still be recycled and the manufactured product could still be remanufactured. Furthermore, the mineral economy could lead to the establishment of a service sector economy in the long run which could exhibit characteristics of 'strong sustainability'. Silicon Valley is an example of a mineral economy from the 19th-century California Gold Rush which eventually led to the conversion of natural capital into financial capital that fuelled a knowledge economy around computing. Tycoons who benefited from the gold rush such as Leland Stanford invested in educational institutions and other enterprises which eventually gave us a service sector economy that no longer relies directly on minerals. Arguably such an economy could not have been possible without the extraction and investment of the natural capital (gold) into other productive forms of capital.

Resource extraction and the quest for circularity

The statistician Harold Hotelling considered primary industries of an economy as the key to a more sustainable economic system. He asked a fundamental question in his research: what should we extract now versus in the future, knowing that the resource is finite? Through a detailed mathematical analysis he showed how we could in fact reach an 'optimal rate of depletion', given various considerations of extraction costs and the price as determined by the number of firms in the market. The key determining factor in his analysis was the discount rate—a measure of how much we value the future compared to the present. The concept was clearly problematic from an environmental perspective, since we were dealing with the tricky issue of inter-generational equity.

Hotelling published a 'rule' in 1931, which posited that 'owners of non-renewable resources will only produce a supply of their basic commodity if it can yield more than available financial instruments, specifically interest-bearing securities'. This rule led to a degree of complacence around pricing and investment returns mechanisms that involved non-renewable resource extraction. Economists such as Nobel laureate Robert Solow and many of his intellectual progeny used this rule to develop dynamic equilibrium models, at the heart of which were key assumptions about *substitutability* of resources. As cost of production increased for one resource, there would be incentives to find substitutes.

The resource economists within the conventional annals of the field largely dismissed any ecological critiques of their models. Hotelling's Rule was subsequently linked to the question of how the profits from exhaustible resource extraction could be linked to creating a sustainable economic trajectory by Canadian economist John Hartwick in 1977. Much of Hartwick's work was focused on considering how resource-rich countries could use the wealth created from natural capital to develop a diversified economy which could reach sustainability. Among environmental studies scholars, the Hartwick–Solow approach to sustainability is referred to as 'weak sustainability' because it ultimately assumes that resource depletion would transition from an extractive to a service-oriented economy.

Using the discount rate makes the daring presumption that parents should be able to estimate the worth of a resource for their children, but even more consequentially for future generations in perpetuity. Yet the reason for having a discount rate is that there is indeed uncertainty for better or for worse in all planning endeavours. The issue of technological progress remains a major unknown. Ideally the depletion rate would somehow keep up with the time lapse in technological development for alternatives, but this is a major assumption.

The question of metal supply became the focus of a famous bet on sustainability of human resource usage between biologist Paul Ehrlich and economist Julian Simon in the 1980s. Ehrlich bet on absolute resource scarcity and price rise of five metals due to population growth and excessive extraction for new technology while Simon bet that we would innovate our way out of the issue. The bet lasted a decade and became a key litmus test for optimism versus pessimism in sustainability science and policy. Simon won the bet and noted his view that even with a finite resource, we should consider human ingenuity and market mechanisms. Grappling with the paradox of unlimited resource extraction from a seemingly finite resource, Simon gave the analogy of the length of a line segment being fixed but the number of points along it being infinite. Simon's metaphor is functionally deceptive in the case of most finite resource concerns, as it relies on the assumption that products would be devisable with ever smaller ingredients of a resource due to innovations. However, his metaphor has greater salience with the particle-level energy harnessing potential of the atom—both in terms of nuclear fission and fusion.

Economists face a challenge when advocating for market mechanisms to address the environmental impacts of resource extraction. The difficulty lies in the fact that markets struggle to incorporate intricate patterns and feedback in environmental systems. This limitation may stem from shortcomings in how human behaviour responds to certain ecological signals. Dasgupta describes this imponderable as follows: 'Markets are able to function well only if the process governing the transformation of goods and services into further goods and services is linear. However, when someone talks of "ecosystem stress" or "ecological thresholds" as ecologists frequently do, they mean states of affairs in systems governed by non-linear processes. For example, market prices may be unable to signal the impending collapse of a local resource base.'

Since natural systems tend to be cyclical, particularly in the biotic context, there is a tendency to link sustainability to circularity. Circularity also suggests greater efficiency, but this is not always the case if energy is the metric by which efficiency is being measured. A key challenge for circularity is to have a replenishable stock of material, but this may mean making some products less durable. A sustainability systems lens on a circular economy must consider such trade-offs as well as the 'rebound effect' of more consumption (discussed in the context of efficiency in Chapter 2).

Within a circular economy model that has better feedback systems the latent demand can be more easily identified. Although the paradox may well hold in the case of materials and energy usage areas with growing demand, it is less likely to be valid in mature markets. For example, in a developing economy energy efficiency and reduced cost may lead people to buy more products to use more energy. However, in a developed market with a reasonably saturated consumption profile (not many more things to add to material and energy demand), efficiency and dematerialization are very valuable.

Ecological approaches to create a more 'renewable' economy

The conventional economic model has been focused on linear material flows from mines to markets with a clear trajectory towards economic growth. However, an ecological economic approach suggests the need to reconfigure the economic systems around natural constraints and have greater circularity in material flows without a predetermined growth agenda. As with any such major shift in human endeavour, a strong philosophical underpinning can help to draw theoretical insights which in turn allow for transferability of concepts across cases. Circular resource flows and a focus on sufficiency would lead to what the doyen of ecological economics, Herman Daly, called a 'steady-state economy'. There were some underlying physical connections between this approach to economics which could be traced back to the Romanian-American economist Nicholas Georgescu-Roegen, who had been a protégé of Joseph Schumpeter—who famously reconciled the sustainability of capitalism as a process of 'creative destruction'. While Schumpeter had noted the biological evolutionary order for capitalism, Georgescu-Roegen focused more on the physical and thermodynamic limits of economic growth. His seminal book *The Entropy Law and the Economic Process* (1971) was the first treatise to consider physical constraints on economic growth. The systems economist Kenneth Boulding was also instrumental in further linking such an approach to social and ecological systems.

Although this approach is clearly well regarded by ecologists, there are two key critiques offered by conventional economists: 'steady state' implies an atrophy of incentives for innovation and hence could diminish the potential for technological advancement of humanity; and the development needs of the poor create a moral imperative for economic growth that would be precluded by a steady-state economy. It is important to note, however, that proponents of ecological economics are willing to embrace growth, so long as material flows are better cycled within the growth paradigm—they are thus focused on stability at the microeconomic level rather than having a steady state at the macroeconomic level.

An important way of operationalizing ecological economics could be to have a more 'circular economy', in which waste flows are reimagined for their use. The concept has gained traction worldwide through the efforts of Dame Ellen MacArthur, who sailed around the world and was inspired to consider a similar circularity for economic systems. In partnership with the World Economic Forum her foundation has developed a series of products to assist businesses and governments configure their efforts in this direction. Figure 23 shows the key features of how waste systems could be reconfigured in such a circular economy across supply chains.

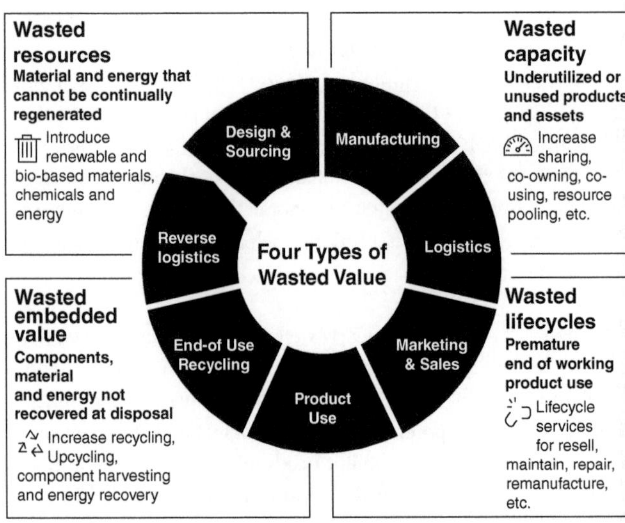

23. Reimagining waste from *The Circular Economy Handbook*.

A neglected aspect of the circular economy discourse has been an evaluation of how such a paradigm would impact basic human development challenges. There seems to be a presumption that 'win-win' outcomes would emerge from efficient systems in a circular economy that could provide development dividends in the world's poorer nations. Yet some of the dominant premises of a circular economy necessitate reduced consumption and increased durability of material products, which has the potential for a major impact on human development in areas that depend on livelihoods from those processes.

Some examples of a circular economy could include: companies offering repair services for electronic devices, extending the lifespan of smartphones or laptops instead of encouraging constant upgrades; or brands creating clothing with easily recyclable materials and designing products with disassembly in mind to facilitate recycling at the end of the product's life; or technologies to convert organic waste into biogas or using waste as a source of energy, thereby extracting value from materials that would otherwise be discarded.

Examining the consumption of various products, services, and the underlying primary resources forms a critical nexus for analysing the intersection of economic development and environmental impact. Scholars have urged for an improved environmental accounting system to meticulously monitor the elemental inputs and outputs. This is essential for evaluating the trade-offs between the positive economic impacts of a project and its detrimental environmental effects. Building upon the classic input–output modelling introduced by Nobel laureate Wassily Leontief, some of his protégés in the field of computer science, particularly Faye Duchin, have further refined these models. A significant concern in adopting an ecological economic model revolves around ensuring continued employment in an economy traditionally based on conventional jobs. Some optimists argue that a shift to a service-oriented sector, generating wealth, could offset the decline in manufacturing employment and livelihoods typical in industrial economies. The transition of jobs after the automation of labour-intensive industries in the past century is often cited in this context. Entrepreneurs play a vital role in fuelling new employment opportunities during such transitions.

A more extreme form of automation that is now gaining massive attention is machine learning, as well as the broader field of Artificial Intelligence (AI). The connection between AI and sustainability is a broad and contested area for research. On the one hand, AI algorithms can help to develop 'smarter' and more efficient cities in terms of energy and resource usage. On the other hand, computing requires vast amounts of energy that is raising concerns about the viability of such tools in large-scale planning ventures.

However, there's a potential downside to embracing a more circular economy. The focus on increasing the value and efficiency of waste materials might displace jobs, particularly for those in less powerful positions. For example, enhancing the value of recyclables may lead to the emergence of recycling companies, potentially impacting waste pickers negatively. Additionally, there are limits to the absorption of employment by the service sector, even in advanced economies.

Exploring the potential for developing countries with large populations to engage in entrepreneurial activities in the green technology sector requires further study, akin to the examination of the Information Technology sector in India. Such analysis would necessitate a global effort to collect and analyse data across material usage supply chains.

In a neoclassical perspective of 'green growth', while increasing efficiency in a more circular economy may create more green jobs in the short and medium term, there could be fewer jobs overall in the long term due to continuous efficiency-driven competition mechanisms. To fully embrace a circular economy, we must consider the spatial dimensions of resource availability and waste management, viewing it through a social-ecological lens.

However, addressing the temporal dimension of resource extraction requires planned conservation and efficient waste cycling. Finding effective energy sources within the bounds of a circular economy is crucial for its long-term viability. A turning point around efficient and effective energy availability could help resolve various contradictions and challenges in reconciling material flows and development.

Critics argue that structured management of material flows might lead to centralized planning, drawing parallels with historical failures. However, the vision of a circular economy, emphasizing intrinsic interrelationships, does not necessarily imply centralized solutions. Instead, a social ecology lens suggests devolved coordination. It is essential to avoid proposing a single set of hierarchically determined solutions and instead focus on global coordination while harnessing planetary synergies. This is particularly crucial when international efforts on development goals could conflict with the most expedient paths towards a circular economy.

When considering ecological economics and the development of a circular economy, it is crucial to maintain a focus on broader systems connections. Several key features of systems thinking can be outlined. Systems are not mere collections of elements; instead, their components are interconnected and mutually influence each other. For instance, a city's transport system involves more than just the sum of buses or trains; it includes how transit nodes interact with each other and with travellers. All components within a system are organized, either by design or through emergent evolution, to fulfil the system's goals. Fish in a shoal exhibit a system behaviour, self-organizing to distract predators—a product of emergent evolution.

Systems can serve specific functions within larger systems, becoming 'subsystems'. Consider a pond with its aquatic plants and animals: it is a self-contained system that also interacts with the local hydrological environment and the atmospheric system, constituting the larger ecosystem. Furthermore, systems incorporate feedback loops, which can be either negative or positive, reinforcing or mitigating trajectories and transitions. An example is found in the climate system, where volcanic eruptions can release sulphur, creating a negative feedback loop for global cooling within what might otherwise be a locally warming event.

Ultimately, these systems-wide efforts will need to be achieved through what is being termed 'Earth Systems Governance'. An international research project supported by the International Science Council and the United Nations University, started in 2009, is aimed at developing a governance architecture for a more sustainable use of natural resources. What is clear from research thus far is the need for governance networks across scales. Such an approach also dovetails with debates on what is termed 'glocalization'—a portmanteau of globalization and localization. This concept emerged from the Japanese concept of *dochakuka*, which means global localization and referred to adaptation of global farming techniques to local conditions.

The Catholic social teaching of 'subsidiarity' suggests that the lowest functional level of governance should be attempted where possible to attain desired outcomes. Generating circular economies at a micro-scale through intentional communities such as eco-villages bears the hallmark of such subsidiarity approaches. Some of the criticism of European Union (EU) policies on a circular economy also emanates from such a detachment from broader planetary policies and linkages, given that the Maastricht treaty which enshrined EU policy mandates was also premised on the principle of subsidiarity. The Church's encyclical on ecology issued by Pope Francis, entitled *Laudato si'* (translated from Medieval

Italian as 'Praise be to You' with subtitle *On Care for Our Common Home*), echoes these themes of subsidiarity as well.

However, there is a clear tension which needs to be acknowledged between the localized push of subsidiarity for sustainable systems versus the pull of globalization as a means of recognizing connections between planetary processes. A conventional reading of social ecology could lead us to ostensibly sustainable communes, meeting their metrics of circularity in terms of waste reuse and regeneration, but with possibly inefficient outcomes or hidden and essential linkages to the global supply chains. The much-quoted aphorism 'think globally, act locally' becomes a futile guide in most contexts because 'thought' and 'action' remain intrinsically connected at multiple scales. For example, to make some communities most effectively self-reliant on energy with wind turbines or solar panels, we still need to harness minerals like terbium or neodymium, which link us to global supply chains. Those linkages are essential for aggregate metrics of sustainability at a planetary level, which need just as much attention as the viability of human communities at a local level.

Chapter 6
Anthropogenic change and sustainability

How might the disjunctures between natural, economic, social, and political orders be reconciled? To do so, it will be fundamenta to highlight the prime dependence of all systems on core natural processes, despite our contention of living in 'the Anthropocene' with an inordinate impact footprint. A rather astounding set of analyses in this vein was conducted by researchers at the Weitzmann Institute in Israel to highlight our impact. By their calculations, in 2020, the total mass of all human-created materia now exceeds all natural biomass. The same team also calculated that 95 per cent of the total biomass of mammals on earth was no either human or animals cultivated by humans for their consumption.

The scale of human impacts on the natural environment in term of material flows has recently been measured for 77 elements fror mining, fossil fuel burning, biomass burning, construction activities, and human apportionment of terrestrial net primary productivity. The latter term refers to the allocation or distributio of the Earth's net primary productivity (NPP)—the amount of energy that plants capture through photosynthesis, minus the energy they expend through respiration—by human activities. It represents the energy available for consumption by other organisms in the ecosystem. This has been compared to natural mass transfer from terrestrial and marine net primary

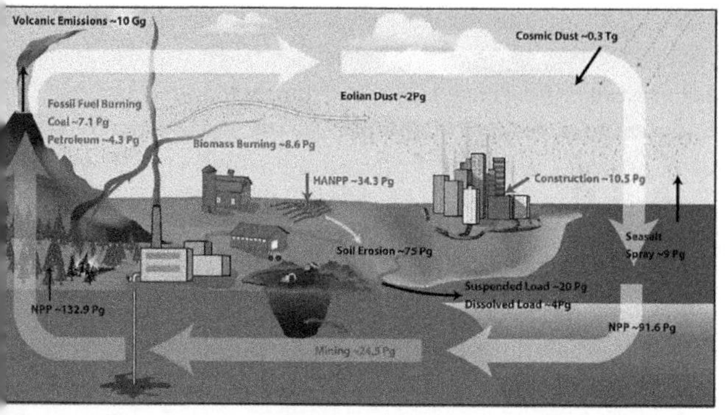

4. Human influence on chemical cycling of the elements. Natural ows include terrestrial and marine net primary productivity (NPP), ea-salt spray, riverine dissolved and suspended particulate latter fluxes to the ocean, cosmic dust, and volcanic emissions. nthropogenic flows include mining, biomass burning, fossil fuel oal and petroleum) combustion, construction activities, and human pportionment of terrestrial net primary productivity (HANPP). lixed natural/anthropogenic flows include soil erosion and aeolian ust. The numbers represent annual flux estimates for the flows. he flux units are Petagram, Pg = 1015g, Teragram, Tg = 1012g, nd Gigagram, Gg = 109g.

roductivity, the transport of dissolved and suspended material ito the ocean, soil erosion, aeolian (wind-blown) dust, sea-salt pray, cosmic dust, volcanic emissions, and, for helium, ydrodynamic escape from the Earth's atmosphere. The findings, ummarized in Figure 24, show that if anthropogenic ontributions to soil erosion and dust are considered, nthropogenic fluxes of up to 62 elements surpass their orresponding natural fluxes.

esearch led by Crelis Rammelt and published in late 2022 onsidered the ways in which the 'Great Acceleration' of human npacts was characterized by a 'Great Inequality' in using and amaging the environment. Remmelt and his team considered hether the notion of 'just access' to minimum needs of energy, water, food, and infrastructure could be met with existing inequalities, technologies, and behaviours. Their research suggests that achieving just access in 2018 in this regard would have produced 2–26 per cent additional impacts on the Earth's natural systems of climate, water, land, and nutrients. These hypothetical impacts, caused by about a third of humanity, equalled those caused by the wealthiest 1–4 per cent. Technological and behavioural changes thus far, while important, did not deliver just access within a stable Earth system. Sustainable societies will need to grapple with social stability just as much as ecological stability as we consider various development trajectories in the Anthropocene.

The dilemma of durability and development

The cover of *Life* magazine in 1955 celebrated the Throwaway Society as a means of liberating the American household. Although there was a shift in behaviour in the 1990s, with the advent of major campaigns such as the 3Rs, 'Reduce, Reuse, Recycle', the COVID pandemic has again revitalized disposability as a virtue. A neglected aspect of transitioning to a durable product paradigm would be its impact on basic human development challenges. There would be jobs lost as product lines are not required to produce greater throughput of disposable products. However, these jobs could be replaced by service sector jobs in repair and remanufacturing of durable products.

There seems to be a presumption that 'win-win' outcomes would emerge from efficient systems in a circular economy that could provide development dividends in the world's poorer nations. Yet some of the dominant premises of a circular economy necessitate reduced consumption and increased durability of material products, which has the potential for a major impact on human development in areas that depend on livelihoods from those processes.

Che consumption of myriad products and services provides an :ssential link between economic development and environmental mpact. Nevertheless, a polarized view that considers consumption ιs only a problem of ecological decadence needs to be avoided.)ptimists in this regard would argue that a transition to a service ector and its concomitant wealth creation would counterbalance he reduced throughput of manufacturing employment and ivelihoods for market economies. However, there are limits to the .bsorption of employment by the service sector in developing ountries with very high populations, such as Nigeria, which are spiring for major development outcomes for their population eyond aggregate indicators of growth. In other words, mere conomic growth data does not always translate to improved uality of life. The latter is often linked to amenities offered by overnments like healthcare and transport infrastructure as well as ɔb satisfaction.

till other environmental thinkers have romanticized subsistence conomies and disparaged entry into markets to prevent eightened consumption or dependence on employment for reating livelihoods. Within the context of a circular economy, it is lso important to note that subsistence has its own challenges hen it comes to realizing the goals of sustainability. Although ubsistence societies have important survival skills for meeting asic human needs for food and shelter, they are so focused on ieir immediate family or tribe that they typically cannot ɔntribute more meaningfully to broader societal ambitions. Thus, ie innovations emanating from subsistence societies are focused n a very narrow sphere of influence. The reduction in nvironmental impact from such insularity which is applauded by iany environmentalists is counterweighted by a possible ınovation deficit that can occur in such cases.

'wenty per cent of consumers say that they replace a device when iere are no more software updates for the old one. The circular :onomy strategy to address this challenge can follow two avenues:

the first would be simply to have a longer warranty support period for the product. The second is to develop better firmware and provide an additional 'keep your device fast' service for existing users for free or for a small upgrade fee. Another software intervention involves enhancing saving opportunities beyond the device through a process known as 'cloud offloading'. This entails offering services that combine access, device, and performance into a bundled package. The primary advantage of cloud offloading is the potential cost reduction, particularly benefiting less established manufacturers. This allows them to concentrate on producing durable, lower-specification new and remanufactured devices instead of directly competing with manufacturers specializing in high-cost, high-specification new devices.

Such efforts could be coupled with better reuse options for old devices which lie abandoned in a drawer—estimates from the Green Alliance's work suggest that one-third of all cell phones ever made are lying in such states of material limbo. The key to maximizing the value and environmental benefits of reuse is to retrieve used devices as early as possible, although there is value in many devices for over five years from the original sale date. For used devices requiring minor repair, the cost of repair is hugely variable: many devices don't have easily removable batteries or replaceable screens, for example, as designers favour slimness over repairability.

Modularity is thus also an important feature of policies that allow for a more sustainable material flow. Remanufacturing and parts harvesting would extend the minor modularity envisaged for improved 'disassemblability' (being able to physically take apart in useful form various components of a product) and compatibility of additional components. This would enable the reuse of components when the device is otherwise not able to be repaired. Such an approach would keep components with high embodied carbon, such as integrated circuits, which contribute 35 per cent of a smartphone's carbon footprint, in use for longer.

Finally, opportunities for 'Do-it-Yourself' repairs are important and gaining traction with sites like 'IFixit'. Redesigning devices for repairability, and providing information on how to repair them, would enable customers to address these problems themselves. However, many IT manufacturers have resisted such an approach: Toshiba has refused to release its repair manuals, citing intellectual property rights, and Apple has developed proprietary screws to prevent customers from opening their devices. Such barriers by industry require regulatory action for change.

Towards the end of 2020, the European Union took some ambitious steps to address the more than 12 million tons of electronic waste the bloc produces annually. Acknowledging that 'Europe is living well beyond planetary boundaries', a European Parliament vote called for mandatory repairability on scores of consumer electronics, amongst a host of other initiatives intended to extend products' life spans. Fortunately, there are also significant advancements in the upgrading of smelters to accommodate e-wastes.

A case in point is the Swedish mining firm Boliden's Rönnskär copper smelter, located outside Skellefteå, which has been recycling various waste materials since the 1960s. To meet the circular economy targets set by the Swedish government, the smelter's annual capacity for recycling electrical material is now 120,000 tons, including circuit boards from computers and mobile phones that are sourced primarily from Europe. The smelted material, known as black copper, then joins the facility's main smelter flow for further refining to extract copper and precious metals.

The smelter has been designed to mitigate the release of potentially hazardous emissions through wet gas purification and other proprietary processes. To abide by the Minamata Convention on mercury pollution, the smelter is also equipped with an additional purification stage for mercury. A key innovation in the design relates to the material–energy nexus: plastic in the electronic material melts during smelting. This molten plastic acts as a source of energy to generate steam in a boiler that is used to drive a turbine which produces electricity or district heating. The heat is partially reused as district heating in the plant area and the remaining heat is supplied to the local district heating system.

With such technologies in place, many nations have made tangible policy interventions—the UK agreed to enforce EU repair rules, France launched a 'repairability index' for select electronics, and Austria reduced taxes on small repairs. COVID-19's impact on consumers and the time being spent at home also appears to have made an impact across the Atlantic, with a particularly inertial US Congress. The Critical Medical Infrastructure Right-to-Repair Act of 2020 marked the first time in US history that a right-to-repair bill was proposed on the national level. Since then, more than half the states in the union have introduced right-to-repair bills that call for equal access to things such as replacement parts, training manuals, and tools. While these bills are often limited to a specific industry—targeting electronics, appliances, automobiles, farming, or medical equipment—the passage of just one could have ripple effects across the nation. In November 2020 Massachusetts passed a resolution to further empower its 2012 automotive right-to-repair law—the first and only right-to-repair law on the books in the USA. The resolution expands the data and diagnostic information automakers are required to provide, thus enabling third-party repairs. Despite its limited scope, the 2012 law led to a national standard for automakers and the recent resolution is expected to have similar effects.

Also noteworthy is the US Department of Energy's launching of the 'Remade Institute', which focuses on accelerating the circular economy through technological solutions. The approach is grounded in five key areas of concentration (Nodes): Systems Analysis and Integration, Design for 'Re-X', Manufacturing Materials Optimization, Remanufacturing and End of Life (EOL) Reuse, and Recycling and Recovery. 'Design for Re-X' refers to a design philosophy that focuses on creating products, systems, or processes with the intention of facilitating reuse, recycling, repurposing, or other forms of resource recovery. The 'Re-X' can represent various actions related to the circular economy, such as redesigning for 'Re-use', 'Re-cycle', 'Re-purpose', or any other 'Re-' term that aligns with sustainable practices.

As an example, aluminium recycling transformed the industry in remarkable ways. In 1960 recycled aluminium accounted for 18 per cent of America's total aluminium supply. Over the next 45 years, production of recycled aluminium rose by almost 746 per cent; during those same 45 years the total US aluminium metal supply increased 300 per cent. The theoretical minimum energy required to produce secondary aluminium at 960 °C is 0.39 kWh per kilogram. So on a theoretical and practical basis the energy required for secondary aluminium is less than 6.5 per cent of the energy required to produce primary metal.

The European Commission has outlined plans to establish a new 'right to repair' for consumers. Currently, EU consumers have a right to have faulty products repaired, but only when a defect is present at the time of delivery and becomes apparent within the legal warranty or guarantee period, which in most EU Member States is two years. The consumer's right to repair could be undertaken using a range of measures: creating a new right to repair for defects caused by wear and tear or mishandling of the product if this arises within a set period—potentially two years, applicable to some categories such as consumer products and electronics; amending existing Sale of Goods Directives to ensure repair, rather than replacement, unless repair is not possible or would be more costly; restarting the legal warranty period for products that have been repaired, or providing consumers with a longer legal warranty period; and by extending the legal warranty period for second-hand and refurbished products to equal that of new products. Currently, the parties can agree to a shorter liability period of not less than one year for second-hand products.

Economists consider such policies as a way to counter the negative effects of 'planned obsolescence'—a term which refers to the producers' incentive that a product should become obsolete after some time, to allow for new sales. This can be motivated by a desire to have the most innovative products or it can also be motivated by a crass desire for throughput. Increasingly, it is becoming apparent that many large corporations are deliberately designing products for planned obsolescence, which is having a massive impact on material flows—especially of metals.

In 2019 alone, by one estimate 50 tonnes of electronic waste were generated globally, with only around 20 per cent of it officially recycled. Half of the waste was from large household appliances and heating and cooling equipment. The remainder was TVs, computers, smartphones, and tablets. The problem has attracted the attention of civil society groups as well. The ability to maintain product life and retain associated jobs and livelihoods requires us to shift from a linear manufacturing model to one coupled with service sector employment in repairs, remanufacturing, and recycling.

Democracy and sustainability

Transitioning to a sustainable but democratic society also needs us to consider consumer choice as an important part of the model, which can't just be 'engineered'. Here too social ecology provides us a way forward, by focusing on a dialectical process of regulation, education, and behavioural change. However, even if regulation is liberally applied to encourage a circular economy or other mechanisms for a more sustainable society, there are still certain fundamental individual liberties that we have now come to accept as beyond the reach of regulation. Regulating birth decisions, for example, would have been the single most potent regulatory mechanism for the ardent neo-Malthusian in mitigating resource depletion and environmental harm, but is no longer plausible as a policy choice. Gone are the days when scholars such as Garrett Hardin were proposing 'Life Boat Ethics' that advocated apathy towards the poor and the elderly or condoned a demise of populations to sustain 'spaceship earth'. Environmentalism has been more universally humanized but that has also led to a conundrum of how best to address our fundamental resource constraints.

Even ardent proponents of population control have mellowed their conversations on the matter, considering the enormous ethical implications of such rash rhetoric. Amartya Sen recognized the challenge of reconciling efficiency and 'optimal' societal behaviour with liberalism several decades ago in his famous essay 'The Impossibility of a Paretian Liberal'. Sen was concerned with human propensity for conflict when certain inalienable values collide within a liberal system that may also be trying to achieve 'Pareto' optimality. Named after the 19th-century Italian economist Vilfredo Pareto, this move towards optimality could be theoretically achieved when a movement from one allocation of resources to another can make at least one individual better off without making any other individual worse off. Sen showed that when we define certain unrestricted domains of human behaviour, such as decisions on having children or what we wear or eat, then we cannot aspire towards having an optimal society.

Kenneth Arrow arrived at a similar insight with regard to voting behaviour in his famous 'impossibility theorem', which suggested that voting systems are not capable of converting the ranked preferences of individuals into a community-wide ranking of societal preferences. Any approach to consumption and the environment must grapple with this fundamental challenge and thus a multifaceted approach with incentive-driven regulations, technological innovation, and literacy-based behavioural change is essential. Human ambition, or what one can refer to as a 'treasure impulse', has been at the heart of human development. This impulse should no longer be limited to a base desire to plunder the Earth's resources. However, in our antipathy towards the negative attributes of human ambition and 'greed' we should not, in my view, diminish the impulse to innovate. Rather, a more responsible channelling of this impulse can encourage extraction practices of minerals that minimize environmental harm, and reduce wastage, as well as facilitate green sources of energy generation. Innovation is the key that leads to the conceptualization of marketable goods and services and the creation of new livelihood opportunities.

To connect these diffuse ideas, Figure 25 shows a framework for how to conceptualize the challenge of achieving a sustainable society in a way that integrates livelihoods around human 'need' (biophysical necessities) and 'greed' (psychosocial attributes that contribute to the quality of life and the ambition to innovate). As the figure illustrates, the crux of my argument is that if people are informed about the impact of their decisions regarding consumption on the environment as well as livelihoods, the goal of sustainability becomes a broader framework within which both the issues of environmental protection and poverty alleviation are included. However, as portrayed in the diagram, these livelihoods can only be sustained by consumption, whether in the case of biological necessities that satisfy 'needs' or social necessities that satisfy 'greed'. Arrows labelled with + signs as well as pathway T indicate potentially positive pathways towards ecological, economic, and social sustainability. Pathway C defines a negative pathway for the same criteria, moving away from sustainability, and all other pathways can have positive or negative outcomes depending on decisions nodes.

Incentive-driven development paths necessitate some measure of consumerism around luxury goods in developed countries.

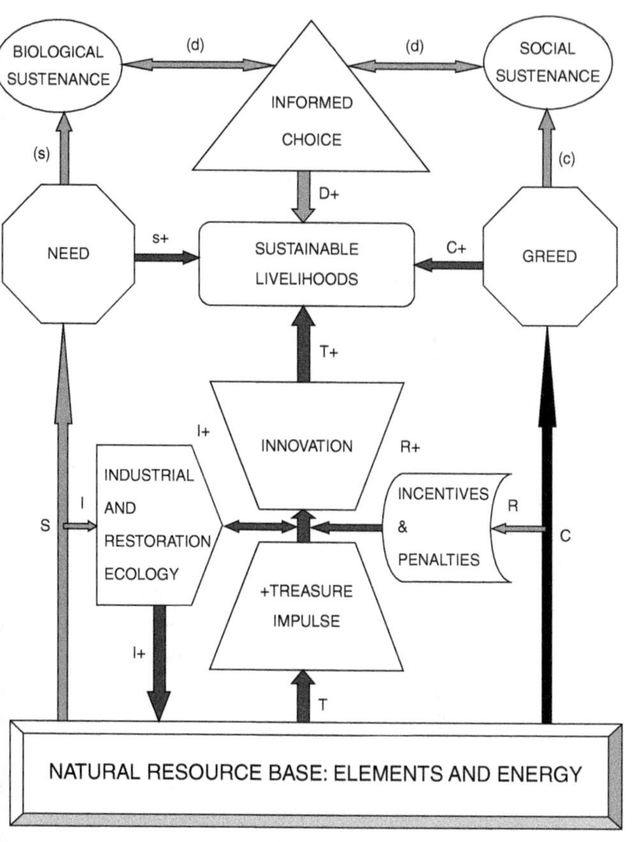

25. Reconciling human 'need' and 'greed' towards sustainability. S = subsistence and survivalist demand; C = direct greed-based consumption (or plunder); R = regulatory measures; I = innovation capital; T = technologically driven demand; D = democratic process. Lower-case notation suggests subsidiary pathway of concept in upper-case. + indicates pathway with definite positive potential for sustainable development.

No doubt the outcome of such a path will be suboptimal, from the perspective of purely environmental conservation. In a society that values some norms of human choice regarding well-being we will always contend with some win-lose propositions. We can educate and regulate but must always be cautious about totalitarianism, for that may stifle our ultimate salvation out of the environmental crisis—the capacity to innovate.

At this stage, those studying public administration may ask whether such innovation should be allowed to happen organically, or whether greater intervention from planners is more efficient. There has been a long-standing debate on this between the 'instrumentalists' such as Charles Lindblom, who wrote a famous essay titled *The Science of Muddling Through*, and planners who argue for 'comprehensive rationality'. While much of human history has been characterized by a level of 'muddling through' when it comes to using resources, if we are to balance multiple objectives over limited time horizons for impact, planning becomes more necessary. However, planning for sustainability does not imply that we merely follow a top-down technocratic route but rather be willing to have a convergent and deliberative process between expertise and the grass roots.

Such an approach is essential to addressing the most vexing question of sustainability: are some negative effects of economic development worth enduring as a necessary sacrifice to reap greater rewards of growth that would self-correct the deleterious impacts of development? This was the prognosis of the work of economist Simon Kuznets, whose name is now immortalized in the famous 'Kuznets curve'. The original curve shows the result of his hypothesis that economic inequality would increase with economic growth but eventually decline. This trend could be explained by structural changes in the economy, such as a shift from agriculture to industry and services, which can contribute to changes in income distribution. Improvements in education and skill levels during the development process may lead to a more equitable distribution of income. Increased urbanization can impact income distribution by altering the dynamics of employment, wages, and access to resources. The same logic was also employed by subsequent economists to environmental harm, suggesting that ecological damage was a price to pay for initial development, after which a self-correcting mechanism would somehow kick in to improve environmental performance. Such an approach is known as the Environmental Kuznets Curve (EKC) hypothesis.

This hypothesis has been debated for at least 25 years in various forms. Many of the controversies have revolved around the scale of the analysis, the kind of environmental concern chosen, and the relative determinism of the pollution reduction with income. The curve also does not account for the pollution haven phenomenon that is associated with growing pockets of unequal pollution impacts. Nemat Shafik, while working at the World Bank, found that for many pollutants the relationship between income and pollution is not shaped like an upside down U (which might suggest that the solution to pollution is more growth) but rather more like a cedilla, 'rising with income, then falling as the low-hanging fruit of pollution abatement is plucked, then rising again as the underlying thermodynamic-physical reality asserts itself'.

However, the empirical evidence has only marginally supported the reduction of inequality and environmental harm with economic development. The literature now suggests that the EKC is by no means deterministic in terms of a development path. There can be variations in its trajectory, based on the pollutant, as well as frequent changes in its inflection depending on what form of development path is chosen. The variation in pollution loading with time needs to be considered over much longer time horizons and also with greater granularity of measurement to gain an accurate understanding of the relationship between economic growth variables and pollution (Figure 26).

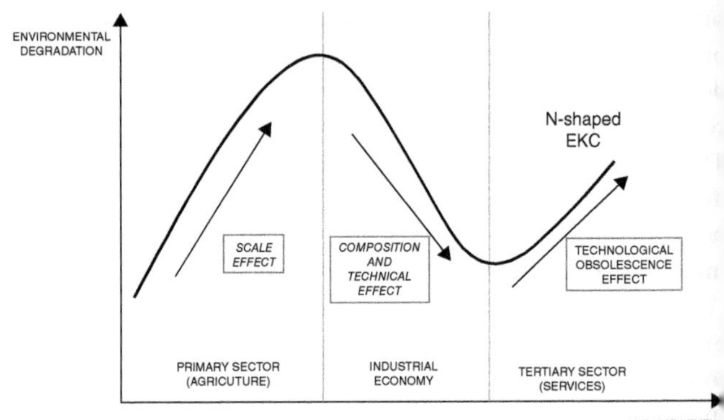

26. Example of an Environmental Kuznets Curve.

The consensus among economists is that the EKC explains some kinds of pollution such as air and noise but not other forms of environmental harm, such as land degradation, deforestation, and soil erosion, particularly in urban ecosystems. An analysis of the Global Environmental Monitoring System (GEMS) database on air and water quality has generally supported the EKC hypothesis and the inflection point of the Kuznets curve for most of the 14 major pollutants tends to occur when the country's annual per capita income reaches around $8,000. However, the analysis also reveals some cautionary results about secondary inflections back towards environmental harm: for example, coliform bacteria, which cause diarrhoea, correlate to per capita income rises with income and then fall, but then rise again beyond a per capita income of $10,000. The inflection points are caused by shifts in the structure of the economy in terms of the kind of production processes and their pollution profile.

The effect of income growth has also been studied with three determinants of pollution: the share of industry in national output, the share of polluting sectors in industrial output, and 'end-of-pipe' (EOP) pollution intensities per unit of output in the polluting sectors. Results indicate that the industry share of national output follows a Kuznets-type trajectory, but the other two determinants do not, and in combination their results implied the rejection of the EKC hypothesis for industrial water pollution. The sectoral composition provides a clean technology dividend for low-income developing countries, but exhibits little or no trend beyond the middle-income range. However, EOP pollution intensity declines continuously with increased income.

Such results point towards a weakness of using econometric techniques in such analysis as well where highly specific variation may be found with certain pollutants and where more qualitative research methods are needed to ascertain any definitive relationship between variables.

The pathway by which economic development can lead to environmental conservation is presented through the EKC in terms of consumer pressure on government to engage in more stringent regulations once a certain income level is achieved, which can then also lead to win-win outcomes of a 'green economy' or 'ecological modernization'. Yet environmental activism is by no means correlated with greater income in and of itself, although in specific cases, it may have greater policy impact in higher-income countries. Moreover, the idea that higher-income groups of countries are more environmentally conscious is also contested, as the poor may be more environmentally friendly than the rich because of their more acute resource reliance.

Furthermore, another important determinant of the EKC can be the influence of trade, whereby pollution intensity in some sectors is simply exported to other parts of the world. Although this may be true for a few sectors like mining of rare earths, which shifted largely to China due to environmental regulations, the most comprehensive evaluation of the embedded pollution of imports suggests that within the USA there has been a gradual shift to greener imports.

Despite the contentions surrounding its empirical observations and the need for a more nuanced approach to pollution policy, th EKC provides a good initial framing mechanism for further unpacking the pollution–development dynamic. Moving from macro-models to specific examples, Malaysia provides an important case study of a country which has shown a rapid increase in development indicators over the past 50 years but ha also fared well on environmental performance indices such as th Yale Environmental Performance Index. However, even in this case of a 'win-win' outcome trajectory, research shows that overa pollution loading, particularly in waterways, has been directly correlated with economic development. Population accounted fo 74 per cent of total polluted rivers and industrial production accounted for 78 per cent of the yearly variances in levels of rive pollution.

An intriguing converse study of the impact of economic contraction and reduced industrial activity on pollution found tha over a 20-year data period in California (1980 to 2000), economi recessions were correlated with reduced pollution. The relationship between the employment measures and air pollutio was statistically significant, suggesting that air quality improves during economic downturns.

Rather than using such macroeconomic back-casting tools, a 'systems engineering' approach to sustainability planning has bee considered more practical and easier to implement in both publi and private sectors. Figure 27 shows such an approach develope at Pennsylvania State University.

One could argue that poverty itself has the potential to generate environmental impacts, and thus, the costs of pollution abateme should be evaluated in the context of how they might exacerbate poverty, leading to a detrimental cycle of de-development. A wide acknowledged causal pathway in this context is the correlation between larger family size and poverty. However, the actual

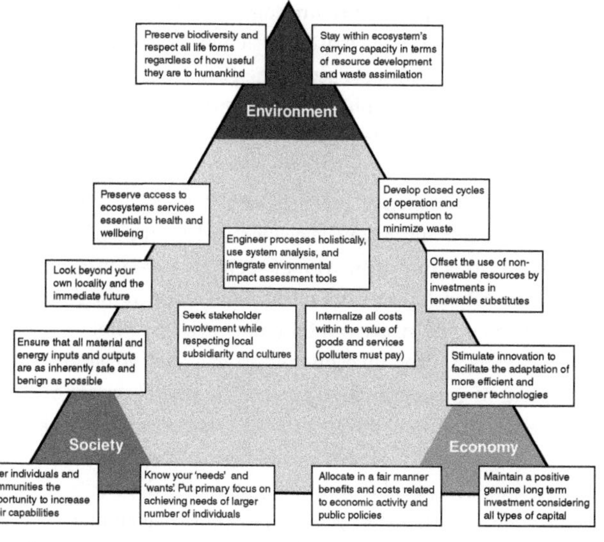

7. A 'systems engineering' approach to sustainability planning.

nvironmental consequences of the relationship between opulation and poverty are a subject of debate.

significant study conducted in Colombia addressed these npacts and concluded that the effects of population and poverty an be effectively influenced by specific policy interventions. For nstance, providing access to low-cost waste management systems an help modulate these impacts, challenging the deterministic ature of the relationship. Additionally, it is crucial to consider the otential 'demographic dividend' associated with a higher opulation, which can contribute to increased labour availability r development and tax income.

he net environmental impact of poverty itself, assessed in the ontext of rural Nepal, was found to be negative but quantitatively egligible: for example, an increase of 10 per cent in income leads to a net fall of 0.2 per cent in firewood collected. The impact of forest degradation (via increased collection times) on local living standards is also minuscule across the Himalayan region, suggesting that demographic factors rather than economic growth itself will determine ecological impacts.

More recently, the idea of 'green growth' has emerged as a policy alternative that could reconcile economic development and pollution. With a mix of smart management and advanced environmental technology, we could avoid many of the deleterious effects of economic growth and use the environmental improvements to prop up economic growth. Indeed, some proponents of green growth suggested that rapid economic growth could help us to tunnel through the EKC and move us quickly to a rich and clean society. However, green growth policies have brought mixed results. For example, green growth policies in South Korea, which was the strongest proponent of the green growth alternative, were questioned as they were based on nuclear energy, construction of dams, and land reclamation, leading to irreversible impacts on the natural environment.

The concept of ecosystem services as a common good that is provided to all strata of society may help to address some of these concerns about environmental injustice as well as providing an accounting mechanism for us to reconcile economic development and environmental conservation. Quantifying the financial value that comes from conserving nature has been a major area for research. This also led to further investment by the international banking community in recent years, including the World Bank, in programmes which can allow for accounting of these ecosystem services. The next question to ask, however, is: if the accounting can be carried out, how might we use financial transactions to help the poor conserve nature? The concept of 'payment for ecosystems services' (PES) has emerged as a result, and is now being widely used as a policy tool to mitigate the ultimate development harms of environmental decline.

Research in this area suggests that the precondition for PES programmes to have beneficial effects on poverty reduction is that the poor should be in the 'right place'—meaning they are in resource-abundant areas; want to participate (e.g. it should 'fit' into the farm practice); and be able to participate (e.g. they should be able to make the necessary investments, have sufficiently secure tenure, etc.). However, studies also conclude that tying together PES and poverty reduction may result in lower efficiency in meeting either objective—and in fact it may be better to focus programmes on one or the other objective separately. Nonetheless, since PES programmes can have indirect effects on the poor—through changes in food prices, wages, and land access—poverty and the poor do need to be taken into onsideration in designing PES programmes, even if poverty eduction is not an objective of the programme.

here is also rising concern about the 'commoditization' of the ecosystem services in a market, which can lead to over-exploitation and evictions of the traditional ecosystem users, to make the ervices available to those who can afford to pay for the ecological ervices. Thus green growth could be achieved, but the benefits ould not be distributed evenly for all. Some level of inequality in erms of aspirations for income level and differences in levels of ontentment is inevitable. However, the level of global evelopmental inequality is staggering, and is itself not sustainable in the context of rising communal violence and social conflict. conomic anthropologist Jason Hickel has in recent years emerged as a prominent voice to link inequality and broader otions of sustainability. He has argued that green growth is himerical in its overall long-term impact, and degrowth is the only able way to mitigate inequality and deal with concerns of cological sustainability. His work is empirically driven, with igorous peer-reviewed publications. In 2022, he published an stimate with collaborators that wealthy countries were esponsible for 74 per cent of global excess resource use over the eriod 1970–2017 and that 'stopping the ecological crisis will require that rich countries pursue transformative post-growth and degrowth policies'. Some of the transformative ideas to decouple 'growth dependencies' would include a reform of the way stock value determines net worth of companies and investment incentives and remuneration of executives.

On the 50th anniversary of the publication of the famous primer on economic sustainability, *Limits to Growth*, by Donella Meadows and other researchers, a 2022 editorial in the journal *Nature* noted that we must 'call time' on the debate around growth The challenge remains how any reform can be accomplished through governance mechanisms which respect some measure of individual choice for consumption. The focus should thus be on tangible steps which need to be taken within these constraints of governance. In some cases, degrowth may well be possible to implement and in others, particularly in developing countries, it would be less plausible. Ultimately, ecological constraints need to be incorporated within economic decision-making for sustainability to be realized.

Sustainability and futurism

Embedded in the concept of sustainability within the Anthropocene is the salience of the future in both its function for humanity and as a biome with a wide diversity of organisms. Sustainable development embodied this imperative through the notion of 'intergenerational equity'—improving the human condition in the present without sacrificing the needs of future generations. In the best-selling novel *Ministry of the Future*, Kim Stanley Robinson describes such a goal through an intergovernmental environmental body. The goal of this ministry is to act as an advocate for the world's future generations of citizens as if their rights are as valid as those of the present generation. Even if this may seem unrealistic as a goal within the incentive structure and immediacy of human behaviour, sustainability requires a certain measure of enchantment about future generations.

A key feature of the futurist approach to sustainability is a recognition that there are many possible visions and scenarios—many of which may be viable for humanity to exist as a species but with varied qualities of life and impacts on the biosphere. None of these futures can be extrapolated from past trends. The pragmatist philosopher José Ortega y Gasset summed up the emergent and complex nature of humanity's quest for sustainability: 'Life is a series of collisions with the future; it is not the sum of what we have been; but what we yearn to be.' The polarized views of sustainability that oscillated between Cornucopian or Cassandran visions are being rejected with a framing around 'science-based targets' for the 'future we want'. International bodies such as the Earth Commission and 'Future Earth' are now defining the conversations in terms of 'just and sustainable transitions' (note the plural form being used for such trajectories).

The same guidance on appreciation for underlying structures should chart our deliberations on 'geoengineering', as we consider stabilization of the Earth's climate and hydrological systems. Milder manifestations of such interventions have been tried through indirect human experiments with alien species introductions worldwide, and even directly through terra-forming experiments. A notable and often neglected experiment of this kind was carried out by Charles Darwin himself after a visit to remote Ascension Island in the South Atlantic. Through the introduction of over 300 species of plant on the relatively barren volcanic island's high slopes, Darwin and Sir Joseph Hooker were able to terraform an ecosystem that is able to capture water from the misty wind flows to this day. A small pond eventually formed and made the island more habitable.

Whether we go all-out for engineering drastic changes through new physical interventions like sending giant mirrors into orbit, chemical injection of cooling aerosols in the upper atmosphere, or more modest 'geomimicry' of geochemical calcium carbonate sequestration we should be guided by our knowledge of those natural parameters over which we are not likely to have control. At the same time, we must strive to change human behaviour to become more aligned with biophysical constraints.

Humanity is striving to find means of keeping track of our sustainability efforts in a variety of ways which remain controversial. The Global Footprint Network is now widely known as an important grass-roots effort to keep track of our ecological impact with rigorous metrics. Individuals can calculate their own footprints online and national and local jurisdictions get an evaluative score. There is also an 'Earth Overshoot' day, which calculates the date when humanity has used more from nature than our planet can renew in the entire year. This date has been receding slowly but during the pandemic moved from July to August, which reflected that under extreme shocks, humanity can reduce consumption dramatically at short notice.

While our consumption has increased again since then, the drive towards sustainability remains strong. Aligning social systems with the laws of nature, as well as having the foresight to selectively develop paths of measured technological intervention, will most likely keep our species within the sustainable scope of a physically habitable world.

References

Chapter 1: Seeking sustainability

Caradonna, J. L. (2016). *Sustainability: A History*. Reprint edition. Oxford: Oxford University Press.

Chertow, Marian R. (2000). 'The IPAT equation and its variants', *Journal of Industrial Ecology*, 4(4), pp. 13–29. <https://doi.org/10.1162/108819800052541927>.

Clark, W. C., and Harley, A. G. (2020). Sustainability Science: Toward a Synthesis. *Annual Review of Environment and Resources*, 45(1), 331–86. <https://doi.org/10.1146/annurev-environ-012420-043621>.

DiNapoli, R. J., Lipo, C. P., and Hunt, T. L. (2021). 'Triumph of the commons: Sustainable community practices on Rapa Nui (Easter Island)', *Sustainability*, 13(21), p. 12118. Available at: <https://doi.org/10.3390/su132112118>.

Fanning, A. L., et al. (2022). 'The social shortfall and ecological overshoot of nations', *Nature Sustainability*, 5(1), pp. 26–36. Available at: <https://doi.org/10.1038/s41893-021-00799-z>.

Motesharrei, S., Rivas, J., and Kalnay, E. (2014). 'Human and nature dynamics (HANDY): Modeling inequality and use of resources in the collapse or sustainability of societies', *Ecological Economics*, 101, pp. 90–102. Available at: <https://doi.org/10.1016/j.ecolecon.2014.02.014>.

Sayre, N. F. (2008). 'The genesis, history, and limits of carrying capacity', *Annals of the Association of American Geographers*, 98(1), pp. 120–34. Available at: <https://doi.org/10.1080/00045600701734356>.

Teicher, J. (2021). Donut Economics has a hole at its core. *Jacobin*, September 24, 2021. Available online https://jacobin.com/2021/09/doughnut-economics-raworth-amsterdam-capitalism-socialism.

Chapter 2: How energy and materials flow through systems

Rosen, M. A., Dincer, I., and Kanoglu, M. (2008). 'Role of exergy in increasing efficiency and sustainability and reducing environmental impact', *Energy Policy*, 36(1), pp. 128–37. Available at: <https://doi.org/10.1016/j.enpol.2007.09.006>.

Valero, A., Valero, A., Calvo, G., and Ortego, A. (2018). 'Material bottlenecks in the future development of green technologies', *Renewable and Sustainable Energy Reviews*, 93, pp. 178–200. <https://doi.org/10.1016/j.rser.2018.05.041>.

Vries, B. J. M. de (2012). *Sustainability Science*. 1st edition. Cambridge: Cambridge University Press.

West, G. (2017). *Scale: The Universal Laws of Growth, Innovation, Sustainability, and the Pace of Life in Organisms, Cities, Economies, and Companies*. London: Penguin.

York, Richard, Rosa, Eugene A., and Dietz, Thomas (2003). 'STIRPAT, IPAT and ImPACT: Analytic tools for unpacking the driving forces of environmental impacts', *Ecological Economics*, 46(3), pp. 51–65. <https://doi.org/10.1016/S0921-8009(03)00188-5>.

Chapter 3: Technological and economic interventions for a sustainable society

Ayres, R. U. (2008). 'Sustainability economics: Where do we stand?', *Ecological Economics*, 67(2), pp. 281–310. Available at: <https://doi.org/10.1016/j.ecolecon.2007.12.009>.

Cardoso, R., Sobhani, A., and Meijers, E. (2022). 'The cities we need: Towards an urbanism guided by human needs satisfaction', *Urba Studies*, 59(13), pp. 2638–59. Available at: <https://doi.org/10.1177/00420980211045571>.

Costanza, R., Fisher, B., Ali, S., Beer, C., Bond, L., Boumans, R., Danigelis, N. L., Dickinson, J., Elliott, C., Farley, J., Gayer, D. E., Glenn, L. M., Hudspeth, T., Mahoney, D., McCahill, L., McIntosh B., Reed, B., Rizvi, S. A. T., Rizzo, D. M., . . . Snapp, R. (2007). 'Quality of life: An approach integrating opportunities, human needs, and subjective well-being', *Ecological Economics*, 61(2), pp. 267–76. <https://doi.org/10.1016/j.ecolecon.2006.02.023>.

Ehrenfeld, J. (2009). *Sustainability by Design*. New Haven: Yale University Press.

Florin, Marie-Valentine, et al., and International Risk Governance Center at EPFL (IRGC), eds. (2023). *Ensuring the Environmental Sustainability of Emerging Technologies*. Lausanne: EPFL. <https://doi.org/10.5075/epfl-irgc-298445>.

Friedman, Benjamin M. (2006). *The Moral Consequences of Economic Growth*. New York: Vintage.

Hamann, M., Biggs, R., and Reyers, B. (2015). 'Mapping social–ecological systems: Identifying "green-loop" and "red-loop" dynamics based on characteristic bundles of ecosystem service use', *Global Environmental Change*, 34, pp. 218–26. Available at: <https://doi.org/10.1016/j.gloenvcha.2015.07.008>.

Kahneman, Daniel, Krueger, Alan B., Schkade, David, Schwarz, Norbert, and Stone, Arthur A. (2006). 'Would you be happier if you were richer? A focusing illusion', *Science*, 312(5782), pp. 1908–10.

Lemisch, Jesse (2001). 'Nader versus the Big Rock Candy Mountain', *New Politics*, 7(3), p. 12.

Miller, Daniel (2001). 'The poverty of morality', *Journal of Consumer Culture*, 1(2), pp. 225–43.

Princen, T. (2005). *The Logic of Sufficiency*. Cambridge, MA: The MIT Press.

Schor, Juliet B. (1999). *The Overspent American: Why We Want What We Don't Need*. New York: Harper Paperbacks.

Scown, Corinne D., and Keasling, Jay D. (2022). 'Sustainable manufacturing with synthetic biology', *Nature Biotechnology*, 40(3), pp. 304–7. <https://doi.org/10.1038/s41587-022-01248-8>.

Twitchell, James B. (2000). *Lead Us into Temptation*. New York: Columbia University Press.

Chapter 4: Tipping points and resilience

Armstrong McKay, D. I., et al. (2022). 'Exceeding 1.5°C global warming could trigger multiple climate tipping points', *Science*, 377(6611), p. eabn7950. Available at: <https://doi.org/10.1126/science.abn7950>.

Farmer, J. D., Hepburn, C., Ives, M. C., Hale, T., Wetzer, T., Mealy, P., Rafaty, R., Srivastav, S., and Way, R. (2019). 'Sensitive intervention points in the post-carbon transition', *Science*, 364(6436), pp. 132–4. <https://doi.org/10.1126/science.aaw7287>.

Franzke, C. L. E., et al. (2022). 'Perspectives on tipping points in integrated models of the natural and human Earth system: Cascading effects and telecoupling', *Environmental Research Letters*, 17(1), p. 015004. Available at: <https://doi.org/10.1088/1748-9326/ac42fd>.

Gatti, L. V., Basso, L. S., Miller, J. B., Gloor, M., Gatti Domingues, L., Cassol, H. L. G., Tejada, G., Aragão, L. E. O. C., Nobre, C., Peters, W., Marani, L., Arai, E., Sanches, A. H., Corrêa, S. M., Anderson, L., Von Randow, C., Correia, C. S. C., Crispim, S. P., and Neves, R. A. L. (2021). 'Amazonia as a carbon source linked to deforestation and climate change', *Nature*, 595(7867), Article 7867. <https://doi.org/10.1038/s41586-021-03629-6>.

Otto, I. M., et al. (2020). 'Social tipping dynamics for stabilizing Earth's climate by 2050', *Proceedings of the National Academy of Sciences*, 117(5), pp. 2354–65. Available at: <https://doi.org/10.1073/pnas.1900577117>.

Rammelt, C. F., et al. (2022). 'Impacts of meeting minimum access on critical earth systems amidst the Great Inequality', *Nature Sustainability*, 6, pp. 1–10. Available at: <https://doi.org/10.1038/s41893-022-00995-5>.

Schelling, T. C. (1978). *Micromotives and Macrobehavior*. W.W. Norton and Company.

Steffen, W., et al. (2020). 'The emergence and evolution of Earth System Science', *Nature Reviews Earth & Environment*, 1(1), pp. 54–63. Available at: <https://doi.org/10.1038/s43017-019-0005-6>.

Wunderling, N., et al. (2022). 'Global warming overshoots increase risks of climate tipping cascades in a network model', *Nature Climate Change*, 13, pp. 1–8. Available at: <https://doi.org/10.1038/s41558-022-01545-9>.

Chapter 5: Renewability, circularity, and industry

Allwood, J. M., et al. (2017). 'Industry 1.61803: The transition to an industry with reduced material demand fit for a low carbon future', *Philosophical Transactions of the Royal Society A: Mathematical, Physical and Engineering Sciences*, 375(2095), p. 20160361. Available at: <https://doi.org/10.1098/rsta.2016.0361>.

Clark, W. C., and Dickson, N. M. (2003). 'Sustainability science: The emerging research program', *Proceedings of the National Academy of Sciences*, 100(14), pp. 8059–61.

Dasgupta, P. (2001). *Human Well-Being and the Natural Environment*. Oxford University Press.

Gibson, R., et al. (2009). *Sustainability Assessment: Criteria and Processes*. London: Earthscan.

Hotelling, H. (1931). 'The Economics of Exhaustible Resources', *Journal of Political Economy*, 39(2), pp. 137–75. <https://doi.org/10.1086/254195>.

Kates, Robert W., et al. (2001). 'Sustainability science', *Science*, 292 (5517), pp. 641–2.

Kenny, Michael, and Meadowcroft, James, eds. (1999). *Planning Sustainability*. London: Routledge.

Markard, J., Geels, F. W., and Raven, R. (2020). 'Challenges in the acceleration of sustainability transitions', *Environmental Research Letters*, 15(8), p. 081001. <https://doi.org/10.1088/1748-9326/ab9468>.

Chapter 6: Anthropogenic change and sustainability

Esty, Daniel C. ed. (2019). *A Better Planet: Forty Big Ideas for a Sustainable Future*. New Haven: Yale University Press.

Gasset, J. O. (1963). *Meditations on Quixote*. New York: W. W. Norton & Company.

George, D. A. R., Lin, B. C., and Chen, Y. (2015). 'A circular economy model of economic growth', *Environmental Modelling & Software*, 73, pp. 60–3. <https://doi.org/10.1016/j.envsoft.2015.06.014>.

Hickel, J. et al. (2022). 'Imperialist appropriation in the world economy: Drain from the global South through unequal exchange, 1990–2015', *Global Environmental Change*, 73, p. 102467.

Kaplinsky, R. (2021). *Sustainable Futures: An Agenda for Action*. 1st edition. Cambridge/Medford, MA: Polity.

Leach, M., et al. (2018). 'Equity and sustainability in the Anthropocene: A social–ecological systems perspective on their intertwined futures', *Global Sustainability*, 1, p. e13. Available at: <https://doi.org/10.1017/sus.2018.12>.

Lenton, T. M., Held, H., Kriegler, E., Hall, J. W., Lucht, W., Rahmstorf, S., and Schellnhuber, H. J. (2008). 'Tipping elements in the Earth's climate system', *Proceedings of the National Academy of Sciences*, 105(6), pp. 1786–93. <https://doi.org/10.1073/pnas.0705414105>.

Mazzocchi, F. (2020). 'A deeper meaning of sustainability: Insights from indigenous knowledge', *The Anthropocene Review*, 7(1), pp. 77–93. Available at: <https://doi.org/10.1177/2053019619898888>.

Shafik, N. (1994). 'Economic development and environmental quality: An econometric analysis', *Oxford Economic Papers*, 46 (Supplement_1), pp. 757–73. <https://doi.org/10.1093/oep/46.Supplement_1.757>.

Swilling, M. (2019). *The Age of Sustainability: Just Transitions in a Complex World.* 1st edition. London: Routledge.

Taylor, G. (2008). *Evolution's Edge: The Coming Collapse and Transformation of Our World.* Gabriola Island, BC: New Society Publishers.

Further reading

Ali, S. H. (2022). *Earthly Order: How Natural Laws Define Human Life*. Oxford: Oxford University Press.

Allenby, Graedel (2009). *Industrial Ecology and Sustainable Engineering*. 1st edition. Hoboken, NJ: Pearson India Education.

Benton-Short, Lisa (2023). *Sustainability and Sustainable Development: An Introduction*. Lanham, MD: Rowman & Littlefield Publishers.

Dietz, Thomas (2023). *Decisions for Sustainability: Facts and Values*. Cambridge, UK: Cambridge University Press.

Esty, D. ed. (2019). *A Better Planet: Forty Big Ideas for a Sustainable Future*. New Haven: Yale University Press.

McDonough, William, Braungart, Michael, and Clinton, Bill (2013). *The Upcycle: Beyond Sustainability—Designing for Abundance*. 1st edition. New York: North Point Press.

Matson, Pamela, Clark, William C., and Andersson, Krister (2016). *Pursuing Sustainability: A Guide to the Science and Practice*. 1st edition. Princeton: Princeton University Press.

Mulligan, Martin (2017). *An Introduction to Sustainability: Environmental, Social and Personal Perspectives*. 2nd edition. London/New York: Routledge.

Portney, Kent E. (2015). *Sustainability*. Illustrated edition. Cambridge, MA: The MIT Press.

Robertson, M. (2017). *Sustainability Principles and Practice*. 2nd edition. London/New York: Routledge.

Sachs, Jeffrey D., and Ki-moon, Ban (2015). *The Age of Sustainable Development*. Illustrated edition. New York: Columbia University Press.

Vries, Bert J. M. de (2012). *Sustainability Science*. 1st edition. New York: Cambridge University Press.

Index

'or the benefit of digital users, indexed terms that span two pages (e.g., 52–53) ay, on occasion, appear on only one of those pages.

A

B

C

L

M

AGRICULTURE

A Very Short Introduction

Paul Brassley and Richard Soffe

Agriculture, one of the oldest human occupations, is practiced worldwide, and is essential for the survival of humanity.

In this *Very Short Introduction*, Paul Brassley and Richard Soffe collate their extensive knowledge on global agriculture, explaining what famers do and why they do it. In regard to controversial issues facing contemporary agriculture, Brassley and Soffe examine agriculture's sustainability, its impact on wildlife and landscape, issues of welfare, and the effect of climate change an the insurgence of genetic modification.

www.oup.com/vsi

BEHAVIOURAL ECONOMICS

A Very Short Introduction

Michelle Baddeley

Traditionally economists have based their economic predictions on the assumption that humans are super-rational creatures, using the information we are given efficiently and generally making selfish decisions that work well for us as individuals. Economists also assume that we're doing the very best we can possibly do—not only for today, but over our whole lifetimes too. Increasingly, however, the study of behavioural economics is revealing that our lives are not that simple. Instead, our decisions are complicated by our own psychology. Each of us makes mistakes every day. We don't always know what's best for us and, even if we do, we might not have the self-control to deliver on our best intentions. We struggle to stay on diets, to get enough exercise, and to manage our money.

This *Very Short Introduction* explores the reasons why we make irrational decisions; how we decide quickly; why we make mistakes in risky situations; our tendency to procrastinate; and how we are affected by social influences, personality, mood, and emotions. As Michelle Baddeley explains, the implications of understanding the rationale for our own financial behaviour are huge. She concludes by looking forward, to see what the future of behavioural economics holds for us.

www.oup.com/vsi

COGNITIVE NEUROSCIENCE

A Very Short Introduction

Richard Passingham

Up to the 1960s, psychology regarded what happened within the mind as scientifically unapproachable. As medical research evolved outlines of brain components and processes began to take shape, and by the end of the 1970s, a new science, cognitive neuroscience, was born.

In this *Very Short Introduction*, distinguished cognitive neuroscientist Richard Passingham gives a provocative and exciting account of the nature and scope of this relatively new field. He explains what brain imaging shows, pointing out common misconceptions, and gives a brief overview of the different aspects of human cognition: perceiving, attending, remembering, reasoning, deciding, and acting. He also considers the exciting advances that may lie ahead.

www.oup.com/vsi

CONSCIENCE

A Very Short Introduction

Paul Strohm

In the West conscience has been relied upon for two thousand years as a judgement that distinguishes right from wrong. It has effortlessly moved through every period division and timeline between the ancient, medieval, and modern. The Romans identified it, the early Christians appropriated it, and Reformation Protestants and loyal Catholics relied upon its advice and admonition. Today it is embraced with equal conviction by non-religious and religious alike. Considering its deep historical roots and exploring what it has meant to successive generations, Paul Strohm highlights why this particularly European concept deserves its reputation as 'one of the prouder Western contributions to human rights and human dignity throughout the world.

www.oup.com/vsi

CORAL REEFS

A Very Short Introduction

Charles Sheppard

Coral reefs have been regarded with awe by millions of people who have encountered them over the centuries. Early seafarers were wary of them, naturalists were confused by them, yet many coastal people benefited greatly from these mysterious rocky structures that grew up to the surface of the sea.

In this *Very Short Introduction*, Charles Sheppard provides an account of what coral reefs are, how they are formed, how they have evolved, and the biological lessons we can learn from them With the vibrancy and diversity of these fascinating ecosystems under threat from overexploitation, Sheppard explores the efforts in place to ensure their conservation.

www.oup.com/vsi

ENVIRONMENTAL LAW

A Very Short Introduction

Elizabeth Fisher

Environmental law is the law concerned with environmental problems. It is a vast area of law that operates from the local to the global, involving a range of different legal and regulatory techniques. In theory, environmental protection is a no brainer. Few people would actively argue for pollution or environmental destruction. Ensuring a clean environment is ethically desirable, and also sensible from a purely self-interested perspective. Yet, in practice, environmental law is a messy and complex business fraught with conflict. Whilst environmental law is often characterized in overly simplistic terms, with a law being seen as be a simple solution to environmental problems, the reality is that creating and maintaining a body of laws to address and avoid problems is not easy, and involves legislators, courts, regulators, and communities.

This *Very Short Introduction* provides an overview of the main features of environmental law, and discusses how environmental law deals with multiple interests, socio-political conflicts, and the limits of knowledge about the environment. Showing how interdependent societies across the world have developed robust and legitimate bodies of law to address environmental problems, Elizabeth Fisher discusses some of the major issues and controversies involved in environmental law.

www.oup.com/vsi

GENOMICS
A Very Short Introduction

John M. Archibald

Genomics has transformed the biological sciences. From epidemiology and medicine to evolution and forensics, the ability to determine an organism's complete genetic makeup has changed the way science is done and the questions that can be asked of it. Its most celebrated achievement was the Human Genome Project, a technologically challenging endeavor that took thousands of scientists around the world 13 years and over 3 billion US dollars to complete.

In this *Very Short Introduction* John Archibald explores the science of genomics and its rapidly expanding toolbox. Sequencing a human genome now takes only a few days and costs as little as $1,000. The genomes of simple bacteria and viruses can be sequenced in a matter of hours on a device that fits in the palm of your hand. The resulting sequences can be used to better understand our biology in health and disease and to 'personalize' medicine. Archibald shows how the field of genomics is on the cusp of another quantum leap; the implications for science and society are profound.

www.oup.com/vsi

GLOBAL WARMING

A Very Short Introduction

Mark Maslin

lobal warming is arguably the most critical and controversial ssue facing the world in the twenty-first century. This *Very hort Introduction* provides a concise and accessible explanation f the key topics in the debate: looking at the predicted impact f climate change, exploring the political controversies of recent ears, and explaining the proposed solutions. Fully updated or 2008, Mark Maslin's compelling account brings the reader ght up to date, describing recent developments from US policy the UK Climate Change Bill, and where we now stand with e Kyoto Protocol. He also includes a chapter on local solutions, eflecting the now widely held view that, to mitigate any npending disaster, governments as well as individuals must act together.

ww.oup.com/vsi

MAMMALS

A Very Short Introduction

T. S. Kemp

From a modest beginning in the form of a little shrew-like, nocturnal, insect eating ancestor that lived 200 million years ago, mammals evolved into the huge variety of different kinds of animals we see today. Many species are still small, and follow the lifestyle of the ancestor, but others have adapted to become large grazers and browsers, like the antelopes, cattle, and elephants, o the lions, hyaenas, and wolves that prey upon them. Yet others evolved to be specialist termite eaters able to dig into the hardest mounds, or tunnel creating burrowers, and a few took to the skie as gliders and bats. Many live partly in the water, such as otters, beavers, and hippos, while whales and dugongs remain permanently in the seas, incapable of ever emerging onto land.

In this *Very Short Introduction*, T. S. Kemp explains how it is a tenfold increase in metabolic rate—endothermy or "warm bloodedness"—that lies behind the high levels of activity, and the relatively huge brain associated with complex, adaptable behaviour that epitomizes mammals. He describes the remarkab fossil record, revealing how and when the mammals gained their characteristics, and the tortuous course of their subsequent evolution, during which many bizarre forms such as sabre-toothe cats, and 30-tonne, 6-m high browsers arose and disappeared. Describing the wonderful adaptations that mammals evolved to suit their varied modes of life, he also looks at those of the mainly arboreal primates that culminated ultimately in Homo sapiens.

www.oup.com/vsi

OCEANS
A Very Short Introduction
Dorrik Stow

The importance of the oceans to life on Earth cannot be overstated. Oceans provide 99% of habitable living space, the largest repository of biomass, and hold the greatest number of undiscovered species on the planet.

In this *Very Short Introduction* Dorrik Stow shows how oceans are an essential resource to our overpopulated world, and discusses why exploration and greater scientific understanding of them is now a high priority. Stow also explores what we know of how oceans originate, evolve, and change, and the inseparable link between oceans and climate.

www.oup.com/vsi

PLANETS
A Very Short Introduction
David A. Rothery

This *Very Short Introduction* looks deep into space and describes the worlds that make up our Solar System: terrestrial planets, giant planets, dwarf planets and various other objects such as satellites (moons), asteroids and Trans-Neptunian objects. It considers how our knowledge has advanced over the centuries, and how it has expanded at a growing rate in recent years. David A. Rothery gives an overview of the origin, nature, and evolution of our Solar System, including the controversial issues of what qualifies as a planet, and what conditions are required for a planetary body to be habitable by life. He looks at rocky planets and the Moon, giant planets and their satellites, and how the surfaces have been sculpted by geology, weather, and impacts.

"The writing style is exceptionally clear and pricise"

Astronomy No

www.oup.com/vsi